Animal Reproduction

Edited by Yusuf Bozkurt
and Mustafa Numan Bucak

Published in London, United Kingdom

IntechOpen

Supporting open minds since 2005

Animal Reproduction
http://dx.doi.org/10.5772/intechopen.94629
Edited by Yusuf Bozkurt and Mustafa Numan Bucak

Contributors
Camila Silva Costa Ferreira, Rita De Cássia Lima Morais, Ştefan Gregore Ciornei, Yung-Sen Huang, Chung-Yen Lin, Kingsley Omogiade Idahor, Mari-Carmen Uribe, Gabino De La Rosa-Cruz, Adriana García-Alarcón, Juan Carlos Campuzano-Caballero, Judith Betsy C, Stephen Sampath Kumar J, Siva C, Feda S. Aljaser, Yusuf Bozkurt, Mustafa Numan Bucak

First published in London, United Kingdom, 2022 by IntechOpen
IntechOpen is the global imprint of INTECHOPEN LIMITED, registered in England and Wales, registration number: 11086078, 5 Princes Gate Court, London, SW7 2QJ, United Kingdom
Printed in Croatia

British Library Cataloguing-in-Publication Data
A catalogue record for this book is available from the British Library

Additional hard and PDF copies can be obtained from orders@intechopen.com

Animal Reproduction
Edited by Yusuf Bozkurt and Mustafa Numan Bucak
p. cm.

This title is part of the Veterinary Medicine and Science Book Series, Volume 11
Topic: Animal Reproductive Biology and Technology
Series Editor: Rita Payan Carreira
Topic Editor: Rosa Maria Lino Neto Pereira

Print ISBN 978-1-83969-514-8
Online ISBN 978-1-83969-515-5
eBook (PDF) ISBN 978-1-83969-516-2
ISSN 2632-0517

For EU product safety concerns:
IN TECH d.o.o., Prolaz Marije Krucifikse Kozulić 3, 51000 Rijeka, Croatia, info@intechopen.com or visit our website at intechopen.com.

IntechOpen Book Series

Veterinary Medicine and Science

Volume 11

Aims and Scope of the Series

Paralleling similar advances in the medical field, astounding advances occurred in Veterinary Medicine and Science in recent decades. These advances have helped foster better support for animal health, more humane animal production, and a better understanding of the physiology of endangered species to improve the assisted reproductive technologies or the pathogenesis of certain diseases, where animals can be used as models for human diseases (like cancer, degenerative diseases or fertility), and even as a guarantee of public health. Bridging Human, Animal, and Environmental health, the holistic and integrative "One Health" concept intimately associates the developments within those fields, projecting its advancements into practice. This book series aims to tackle various animal-related medicine and sciences fields, providing thematic volumes consisting of high-quality significant research directed to researchers and postgraduates. It aims to give us a glimpse into the new accomplishments in the Veterinary Medicine and Science field. By addressing hot topics in veterinary sciences, we aim to gather authoritative texts within each issue of this series, providing in-depth overviews and analysis for graduates, academics, and practitioners and foreseeing a deeper understanding of the subject. Forthcoming texts, written and edited by experienced researchers from both industry and academia, will also discuss scientific challenges faced today in Veterinary Medicine and Science. In brief, we hope that books in this series will provide accessible references for those interested or working in this field and encourage learning in a range of different topics.

Meet the Series Editor

Rita Payan Carreira earned her Veterinary Degree from the Faculty of Veterinary Medicine in Lisbon, Portugal, in 1985. She obtained her Ph.D. in Veterinary Sciences from the University of Trás-os-Montes e Alto Douro, Portugal. After almost 32 years of teaching at the University of Trás-os-Montes and Alto Douro, she recently moved to the University of Évora, Department of Veterinary Medicine, where she teaches in the field of Animal Reproduction and Clinics. Her primary research areas include the molecular markers of the endometrial cycle and the embryo–maternal interaction, including oxidative stress and the reproductive physiology and disorders of sexual development, besides the molecular determinants of male and female fertility. She often supervises students preparing their master's or doctoral theses. She is also a frequent referee for various journals.

Meet the Volume Editors

Yusuf Bozkurt has a BSc, MSc, and Ph.D. from Ankara University, Turkey. He is currently a Professor of Biotechnology of Reproduction in the field of Aquaculture, İskenderun Technical University, Turkey. His research interests include reproductive biology and biotechnology with an emphasis on cryo-conservation. He is on the editorial board of several international peer-reviewed journals and has published many papers. Additionally, he has participated in many international and national congresses, seminars, and workshops with oral and poster presentations. He is an active member of many local and international organizations.

Mustafa Numan Bucak received a bachelor's degree from the Veterinary Faculty, Ankara University, Turkey, where he also obtained a Ph.D. in Sperm Cryobiology. He is an academic staff member of the Department of Reproduction and Artificial Insemination, Selçuk University, Turkey. He manages several studies on sperms and embryos and is an editorial board member for several international journals. His studies include sperm cryobiology, in vitro fertilization, and embryo production in animals.

Contents

Preface

Today, global food protein demand is moving from plant-based proteins to animal-based proteins. It is expected that the requirement for animal-based foods will soon almost double. As such, enhancing the reproduction potential of animals and applying reproductive biotechnological techniques have great importance in the sustainability of terrestrial and aquatic animal production.

Reproductive biotechnology is a technology that develops animal production by improving the reproductive inadequacy of animals by different techniques. Nowadays, different assisted-reproductive tools have been developing and evolving to obtain many offspring from genetically superior terrestrial and aquatic animals. Hence, this book highlights the reproductive strategies and important practices in the reproduction of terrestrial and aquatic animals. It is divided into five sections and eight chapters written by experts in their specialized fields.

Section 1, "Overview of Animal Reproduction", provides a snapshot of reproduction strategies and biotechnologies in terrestrial and aquatic animals.

Section 2, "Reproductive Techniques in Terrestrial Animals", includes three chapters. The first chapter, "Cryopreservation Methods and Frontiers in the Art of Freezing Life in Animal Models", emphasizes the reasons for the conservation of animal genetic resources and cryopreservation biotechnology with all aspects in terrestrial animals. The second chapter, "Embryo Transfer", describes the methodology of embryo transfer in terrestrial animals. The third chapter, "Doppler Ultrasound in the Reproduction of Mares", covers the examination of reproductive systems via ultrasonography with emphasis on blood flow in mares.

Section 3, "Reproductive Techniques in Aquatic Animals", includes one chapter entitled "Cryopreservation and Its Application in Aquaculture," which describes the importance of aquaculture and the structure of cryopreservation biotechnology for aquatic animals.

Section 4, "Reproduction in Avian Species", includes one chapter entitled "Avian Reproduction," which describes avian species and their reproduction peculiarities.

Section 5 includes two chapters. The first chapter, "Stimulatory Effects of Androgens on Eel Primary Ovarian Development - from Phenotypes to Genotypes", covers artificial maturation and ovarian development by applying androgen as a stimulator. The second chapter, "Intraovarian Gestation in Viviparous Teleosts: Unique Type of Gestation Among Vertebrates", debates the structure of intraovarian gestation in viviparous teleosts.

This book addresses different topics regarding the reproduction of terrestrial and aquatic animals. It is a useful resource for researchers studying in this field.

We would like to thank all the authors for their distinguished contributions, as well as Author Service Manager Ms. Paula Gavran at IntechOpen for her cooperation at various phases of the publication process.

Yusuf Bozkurt
İskenderun Technical University,
İskenderun, Hatay, Türkiye

Mustafa Numan Bucak
Selçuk University,
Konya, Türkiye

Section 1

Overview of Animal Reproduction

Chapter 1

Introductory Chapter: Reproductive Strategies and Biotechnologies in Terrestrial and Aquatic Animals

Yusuf Bozkurt and Mustafa Numan Bucak

1. Introduction

Reproduction is a biological process by which new individual organisms are produced from their parents. Animal reproduction aims to renew generations for a given production purpose such as meat, milk, or wool according to species. In order to achieve this goal researchers look for the best control of reproduction in parents to provide a maximum number of new-born of the required quality [1].

Nowadays, many reproductive biotechnologies have been developed for the effective control of reproduction in terrestrial and aquatic animals showing great differences in reproduction types. The application of biotechnology offers many advantages to animal food production through enhancement and control of reproductive processes in terrestrial and aquatic animals.

2. Reproductive strategies

Reproduction success is a very important target for all animal species. Animal species take different strategies to achieve this target [2]. The objective of a reproductive strategy is to maximize reproductively active offspring depending on available energy and parental life expectancy [3].

There are two forms of reproduction known as asexual and sexual. There are several ways of reproducing asexually such as fission, budding, fragmentation and parthenogenesis. Asexual reproduction occurs in prokaryotic microorganisms (bacteria and archaea) and many eukaryotic, single-celled, and multi-celled organisms.

Additionally, some animal species (including sea stars and sea anemones, as well as some insects, reptiles, and fish) are capable of asexual reproduction. The most common forms of asexual reproduction for stationary aquatic animals include budding and fragmentation where part of a parent individual can separate and grow into a new organism.

On the other hand, sexual reproduction typically depends on the sexual interaction of two specialized organisms, known as gametes, which contain a half number (n) of chromosomes of normal cells and are created via meiosis. Typically, a male fertilizes a female of the same species to create a zygote (2n). In this way, it is possible to produce offsprings whose genetic characteristics are derived from those of the two parental organisms.

After fertilization, a series of developmental stages occur in which primary germ layers are established and organized to form an embryo. During this process, animal tissues begin to organize into organs to determine their future morphology and physiology.

It is known that all terrestrial animals perform internal fertilization, whereas aquatic animals perform various reproductive systems including internal fertilization with or without mating. On the other hand, many aquatic animals perform mainly external fertilization in different types such as viviparous, oviparous, and parthenogenesis [4].

2.1 Reproductive strategies in terrestrial animals

Terrestrial animals mostly perform internal fertilization and there are two ways of producing offspring via this way which are:

The first one is fertilizing of the eggs inside of the female's body and the embryo receives nourishment from the egg yolk. When the offsprings are fully developed they are hatched.

The second one is fertilizing of the eggs inside of the female and the embryo receives nourishment from the mother's blood through a placenta. The offspring develops inside of the female body and is born alive.

By the way of internal fertilization, the embryo is isolated within the female and is provided protection against the predators, which also increases the likelihood of fertilization of the egg by a specific male. Despite fewer offspring are produced via this fertilization type, but their survival rates are higher than that of external fertilization.

2.2 Reproductive strategies in aquatic animals

Aquatic animals mostly perform external fertilization. In general, this type of fertilization occurs in aquatic environments where both eggs and sperm are released into the water. The role of water is to protect the eggs from drying out during embryonic development. The gametes are released at the same time to the same location and provide increasing in the likelihood of fertilization of the eggs eventually. The embryo receives nourishment from the egg yolk.

Additionally, there are some strategies used by fish to ensure their offspring survive in aquatic animals. The reproductive strategies of aquatic animals are often reflected in the anatomical differences between the sexes. In this framework, aquatic animals, mainly teleost fishes, have developed a large variety of reproductive strategies, varying from mass spawning to parental care, from strict gonochorism to hermaphrodism, and from oviparity to viviparity.

2.2.1 Gonochorism or hermaphrodism

Most of the fish species are gonochoric, that is their sexes are separate during their lifetime. However, many fish species exhibit sex change that is called as hermaphrodism, which can be protandrous if the fish sex changes from male to female. Other hermaphrodite species can be protogynous hermaphrodite and their sex change from female into a male. Interestingly, simultaneous hermaphrodites also exist in aquatic animals, which behave almost simultaneously as males or females. In this way, individuals can change within minutes from displaying male sexual behavior, with sperm release, to female sexual behavior, with egg-laying [5].

2.2.2 Oviparity or viviparity

In oviparous fish, eggs are fertilized externally, after spawning. On the other hand, in viviparous fish, such as the guppy (*Poecilia reticulata*) or the mosquitofish (*Gambusia affinis*), the fertilized and developing eggs remain inside of the female body for a certain period and the offsprings are released as free-swimming larvae [5].

3. Reproductive biotechnologies

Biotechnology can be defined as a technique using living organisms to modify or improve products. Biotechnology has a great impact on species improvement, reproductive rate, and animal production. Reproductive biotechnology is modern technological technique using biological systems and organisms to develop and increase the quality of the products. The most common reproductive applications integrated with biotechnology are artificial insemination, short and long-term preservation of sperm, sperm sexing, synchronization, superovulation, embryo transfer, and *in vitro* embryo production [6].

3.1 Reproductive biotechnologies in terrestrial animals

The application of biotechnology offers numerous advantages to livestock production through control of the reproductive process in animals. These biotechnologies can be summarized as artificial insemination, estrus synchronization, embryo transfer, sperm cryopreservation, transgenesis, and *in vitro* fertilization. Theuse of these reproductive biotechnologies has contributed tremendously to meet the increasing demands of the modern dairy and beef industry [7].

3.1.1 Artificial insemination

Artificial insemination is one of the earliest reproductive biotechnologies and permits using of superior males for breeding purposes. It involves sperm collection from superior males, dilution of sperm, freezing of sperm, and transferring of frozen–thawed sperm to the female reproductive tract [7].

3.1.2 Estrus synchronization

Estrus synchronization is a grouping of females for parturition at the same time. It is used at commercial dairy farms for uniform milk production throughout the year. It is closely linked with artificial insemination and is also the first step of embryo transfer. Using estrus synchronization is a good strategy to overcome breeding problems, especially during the summer months [7].

3.1.3 Embryo transfer

Embryo transfer is a process by which embryos are collected from a donor female and then transferred into a recipient female where the embryos complete their development. It is the most commonly used biotechnology after artificial insemination and estrus synchronization. Embryo transfer is profitable for producers of pure-bred animals and genetically superior females that produce more offspring than natural reproduction. This technique is used in several species of domestic animals including cows, horses, goats, and sheep [7].

3.1.4 Sperm cryopreservation

Research on sperm cryopreservation can be traced back to the discovery of the protective peculiarity of glycerol for freezing of avian sperm by Polge et al. [8]. Since then, cryopreserved sperm of livestock has grown into a near-billion dollar global industry. Additionally, since the first successful cryopreservation of bull semen, cryopreserved sperm has been used to propagate the rare and endangered species [9].

It is clear that sperm cryopreservation contributes to the expansion of reproductive techniques, such as artificial insemination and *in vitro* fertilization [10]. Additionally, sperm banks are currently more developed for rare domestic breeds (cattle, sheep, and goats) than for non-domestic species. Nowadays, using of sperm banks to facilitate management and conservation of endangered species is being promoted extensively [11].

3.1.5 In vitro fertilization

In vitro fertilization (IVF) is the collection of oocytes from a donor female and fertilization of that matured oocyte in a laboratory dish. The eggs after collection are placed in CO_2 incubators in the IVF laboratory. Most viable spermatozoa are recovered after processing for inseminating the eggs. The addition of a large number of viable spermatozoa to each ova will disperse the follicular cells and also ensure fertilization of an egg by one spermatozoon. *In vitro* fertilization has been used to treat many infertility issues such as when both fallopian tubes are blocked and compulsory fertilization of the egg cell has to take place outside of the body [7].

3.2 Reproductive biotechnologies in aquatic animals

Successful reproduction of cultured fish broodstock is essential in terms of the sustainable aquaculture of aquatic organisms. In this concept, reproductive biotechnologies mainly include cryopreservation of male gametes, genetic control of sex, and production of transgenic fish.

3.2.1 Cryopreservation of male gametes

Cryopreservation is the process of freezing the biological materials at the temperature of liquid nitrogen (LN_2) (−196°C). In this way, it is possible to storing the biological materials as unchanged for centuries with the capability of recovering the cell functionality following the thawing process [12]. Today, sperm management techniques have been established for freshwater and marine fish species [13–15].

Cryopreservation of male gametes is an important biotechnological tool for aquatic species and has great concern for aquaculture. Following successful cryopreservation of avian spermatozoa using glycerol as cryoprotectant by Polge et al. [8], cryopreservation of male gametes became possible in aquatic animals [9]. First time, Blaxter [16] applied a similar approach for fish sperm and reported achieving approximately 80% cellular motility following thawing of Atlantic herring sperm cells. Since then, cryopreservation of fish sperm has been studied and succeeded in more than 200 species [17]. In addition, it is possible to reconstruct the original strain, population, or variety following required environmental restoration via this biotechnology [9].

3.2.2 Genetic control of sex

The genetic control of fish sex could be useful where one sex displays advantageous characteristics such as larger adult size, production of high-value caviar, faster growth rate, or higher age at first sexual maturation. In this concept, monosex populations of the most advantageous sex may be produced through genetic control (crossing, gynogenesis, androgenesis, hybridization) or steroid treatment of broodstock (hormonal treatment) [18].

Genetic control of sex is possible via crossing, gynogenesis, androgenesis, and hybridization. In this framework, it is possible to control sex by the crossing of sex-reversed adult broodstock (administering androgens to produce "neo" males and estrogens to produce "neo" females) with normal males or females to produce single-sex progeny. Another sex control method is gynogenesis which is the fertilization of oocytes with inactivated sperm (irradiated) with normal sperm, to eliminate the female chromosomes and produce all-female progeny. Also, it is possible via androgenesis that is fertilization of inactivated oocytes (irradiated) with normal sperm, to eliminate the female chromosomes and produce all-male progeny. Additionally, another way of genetic sex control is the hybridization of two different species from the same genera producing single-sex progeny [18].

3.2.3 Production of transgenic fish

The domestic fish production through transgenic techniques offers many potential economic advantages for commercial aquaculture production. The traditional method of producing transgenic fish is still microinjection. However, some success has also been shown using particle bombardment [18].

4. Conclusion

Reproduction is the backbone of animal production and productivity is the key element for development. Reproductive inefficiency is one of the most important reasons of economic losses in animal industries. Despite the remarkable advancement in the field of reproductive physiology, low conception rate, and the high embryonic mortality rate remains a major problem in terrestrial and aquatic animals as well.

Author details

Yusuf Bozkurt[1]* and Mustafa Numan Bucak[2]

1 Faculty of Marine Sciences and Technology, İskenderun Technical University, İskenderun, Hatay, Turkey

2 Faculty of Veterinary Medicine, Selçuk University, Konya, Turkey

*Address all correspondence to: yfbozkurt@hotmail.com

References

[1] Courot M, Volland-Nail P. Management of reproduction in farm animals: Present and future. Biotechnology & Biotechnological Equipment. 1991;**5**(3):3-12

[2] Balon EK. Patterns in the evolution of reproductive styles in fishes. In: Potts GW, Wootton RJ, editors. Fish Reproduction: Strategies and Tactics. New York: Academic Press; 1984. pp. 35-53

[3] Roff DA. The Evolution of Life Histories: Theory and Analysis. New York, USA: Chapman and Hall; 1992. p. 535

[4] Yoshida M. Reproductive systems in aquatic animals. In: Yoshida M, Asturiano J, editors. Reproduction in Aquatic Animals. Springer: Singapore; 2020. pp. 13-21

[5] Biran J, Levavi-Sivan B. Endocrin control of reproduction, fish. Encyclopedia of Reproduction (Second Edition). 2018;**6**:362-368

[6] Qamar AY, Khan AU, Jamil A, Abubakar M. Biotechnology and animal reproduction. In: Abubakar M, Saeed A, Kul O, editors. The Role of Biotechnology in Improvement of Livestock. Berlin Heidelberg: Springer-Verlag; 2015. pp. 1-26

[7] Hadgu A, Fesseha H. Reproductive biotechnology options for improving livestock production: A review. Advances In Food Technology and Nutritional Sciences – Open Journal. 2020;**6**(1):13-20

[8] Polge C, Smith A, Parkes A. Revival of spermatozoa after vitrification and dehydration at low temperatures. Nature. 1949;**164**:666

[9] Bozkurt Y. The Role of Cryobiology in Conservation of Aquatic Genetic Resources. Saarbrücken, Germany: Lambert Academic Publications; 2017. p. 94

[10] Medeiros CMO, Forell F, Oliveira ATD, Rodrigues JL. Current status of sperm cryopreservation: Why isn't it better? Theriogenology. 2002;**57**(1):327-344

[11] Wildt DE, Pukazhenthi BS, Brown JL, Monfort S, Howard JG, Roth TL. Spermatology for understanding, managing and conserving rare species. Reproduction, Fertility, and Development. 1995;**7**:811-824

[12] Bozkurt Y. Introductory chapter: Application fields of cryopreservation biotechnology. In: Bozkurt Y, editor. Cryopreservation Biotechnology in Biomedical and Biological Sciences. London, UK: IntechOpen Publishing; 2018. pp. 1-4

[13] Yavas I, Bozkurt Y. Effect of different thawing rates on motility and fertilizing capacity of cryopreserved grass carp (*Ctenopharyngodon idella*) sperm. Biotechnology and Biotechnological Equipment. 2011;**25**(1):2254-2257

[14] Bozkurt Y, Yavas I, Yıldız C. Effect of different avian egg yolk types on fertilization ability of cryopreserved common carp (*Cyprinus carpio*) spermatozoa. Aquaculture International. 2014;**22**:131-139

[15] Bozkurt Y, Yavas I, Bucak MN, Yeni D. Effect of different cryoprotectants (glycerol, methanol and Dimethylsulfoxide) on post-thaw quality, viability, fertilization ability and DNA damage of cryopreserved Nile tilapia (*Oreochromis niloticus*) spermatozoa. CryoLetters. 2019;**40**(1):11-17

[16] Blaxter J. Sperm storage and cross-fertilization of spring and autumn

spawning herring. Nature. 1953;**172**: 1189-1190

[17] Tiersch TR, Yang H, Jenkins JA, Dong Q. Sperm cryopreservation in fish and shellfish. Society of Reproduction and Fertility Supplement. 2007;**65**: 493-508

[18] Muir WM, Hostetler HA. Transgenic fish: Production, testing, and risk assessment. In: Renaville R, Burny A, editors. Biotechnology in Animal Husbandry. Dordrecht, Netherlands: Kluwer Academic Publishers; 2001. pp. 261-281

Section 2

Reproductive Techniques in Terrestrial Species

Chapter 2

Cryopreservation Methods and Frontiers in the Art of Freezing Life in Animal Models

Feda S. Aljaser

Abstract

The development in cryobiology in animal breeding had revolutionized the field of reproductive medicine. The main objective to preserve animal germplasm stems from variety of reasons such as conservation of endangered animal species, animal diversity, and an increased demand of animal models and/or genetically modified animals for research involving animal and human diseases. Cryopreservation has emerged as promising technique for fertility preservation and assisted reproduction techniques (ART) for production of animal breeds and genetically engineered animal species for research. Slow rate freezing and rapid freezing/vitrification are the two main methods of cryopreservation. Slow freezing is characterized by the phase transition (liquid turning into solid) when reducing the temperature below freezing point. Vitrification, on the other hand, is a phenomenon in which liquid solidifies without the formation of ice crystals, thus the process is referred to as a glass transition or ice-free cryopreservation. The vitrification protocol applies high concentrations of cryoprotective agents (CPA) used to avoid cryoinjury. This chapter provides a brief overview of fundamentals of cryopreservation and established methods adopted in cryopreservation. Strategies involved in cryopreserving germ cells (sperm and egg freezing) are included in this chapter. Last section describes the frontiers and advancement of cryopreservation in some of the important animal models like rodents (mouse and rats) and in few large animals (sheep, cow etc).

Keywords: cryopreservation, fertility, reproduction, cryobiology

1. Introduction

Ever since the human evolution and civilization, human has been exploiting animals either for food, transport and or as companions. The use of animals in biomedical and behavioral research has greatly increased scientific knowledge and has benefitted human health enormously. Tremendous advancement took place in the field of medical sciences with the usability of animals for experimental research. Currently, around 75–100 million vertebrates are used annually in research and testing [1]. The most frequently used animals are mice and rats that constitute approximately 95% of experimental animals; mouse being the most commonly used animal in biomedical research [1]. Animal models are used in research for wider understanding of vital physiological processes in human and animals. Animal models are also useful in investigating various diseases including metabolic disorders such as diabetes, cardiovascular disease (CVD), disorders in reproductive endocrinology, infertility, cancer and infectious diseases [2–7]. Human lives endangered due to organ failure were restored after successful organ transplantation accomplished in animal models. Optimization of cryopreservation protocols will significantly facilitate organ transplantation and/or replacement. In addition, animals like dogs, pigs, cats, sheep, non-human primates (NHP) and fish are widely used for genetic and physiological studies in human health and disease [8]. The chapter is focused on the fundamentals of cryobiology and strategies in male (sperm) and female fertility (oocyte and ovarian tissue) cryopreservation. The last section is focused on the frontiers in cryopreservation of most widely used animal models like rodents and higher animals used in biomedical research and toxicological studies.

1.1 Necessity to cryopreserve animal's genetic resources?

Large numbers of animal breeds worldwide are either extinct or endangered with few at the verge of extinction. Hence, it is crucial to develop and apply rescue strategies to ensure survival of these species for the future. One way is to preserve the genetic resources or the germplasm of these species for their maintenance and future development. Genetic diversity is another threat resulting from animal husbandry errors that can result in genetic drift of existing colonies and genetic contamination of lines. Furthermore, weather related natural disaster is a major threat to animal husbandry and vivarium, especially in case of experimental animals that require special care in breeding and rearing colonies. Cattle and breeding industries are modified for the large-scale production of the animal species. There is an urgent need to improve the efficiency and sustainability of producing animals for food in the face of the ever-increasing world population. Improved understanding of mechanisms and challenges of reproductive technologies are vital for improving the viability of the livestock industry [9]. Hence, one solution to preserve animal species is by freezing. Cryopreservation has emerged as the most efficient and compatible method for freezing human and animal genetic resources [10]. Cryobiology is an integrated study of various biological and physical sciences. Semen, embryos, oocytes, somatic cells, nuclear DNA, and other types of biological material such as blood, serum, and tissue, can be stored using cryopreservation, in order to preserve genetic materials or for other applications [8, 11]. The primary benefit of cryopreservation is the ability to save germplasms for extended periods of time, and thus maintaining the genetic diversity of a species or a breed [12]. Also, germplasm of genetically engineered animals (GEA) and cell lines of various species can be preserved by cryopreservation method. Gene banks/cryobanks are established and contain repository of cryopreserved genetic resources to regenerate a particular population in the future [13–15]. Sperm cryopreservation has been successfully applied in various fields to benefit the mankind and animals. Of prime significance, assisted reproduction technology (ART), the forefront in infertility treatment today might be inconceivable without the efficient cryopreservation techniques.

1.2 History behind the discovery

Spallanzani's observation in the beginning of eighteenth century snow cooling the sperm was a breakthrough in the field of biology. Sperm was found to maintain motility and viability even when exposed to cold temperature conditions [16]. Later, in the nineteenth century a major breakthrough occurred with the successful cattle insemination performed using cryopreserved samples. A successful protocol for sperm freezing and storage at low temperatures (−79°C) was developed by Polge *et al.,* [17]. Spermatozoon was the first mammalian cell to be successfully frozen [18]. This significant success in sperm freezing was associated with the discovery of glycerol as a cryoprotectant by Polge *et al* in 1949 [17]. Since 1970 and till date, cryopreservation would not be successful preservation procedure in reproductive medicine without the discovery of plastic straws and cryotubes that tolerate extreme low temperature used in packaging sperm. Mouse embryo freezing was reported using dimethylsulfoxide (DMSO) as a cryoprotectant in 1972 after successfully cryopreserving spermatozoa [19].

2. Art of freezing life

2.1 Cryopreservation methods

Method of cryopreservation and recovery involves following steps. Initially, cells are treated with cryoprotective agents (CPAs) to prevent cryoinjury/damage. Later, cells are cooled in a controlled manner at subzero temperatures, at which the metabolic processes of the cell stops. The recovery of the cells follows a reversed procedure; cells are rapidly thawed and then the CPAs are gradually removed. The viability of the cryopreserved sample is enhanced with the use of appropriate CPA type and concentration and the appropriate rate of freezing [10].

The two basic methods of cryopreservation include the conventional slow freezing (SF) and the rapid freezing method, known as vitrification (Vit). These protocols require different concentrations of CPAs and apply different cooling rates. **Figure 1** illustrates the two methods of cryopreservation.

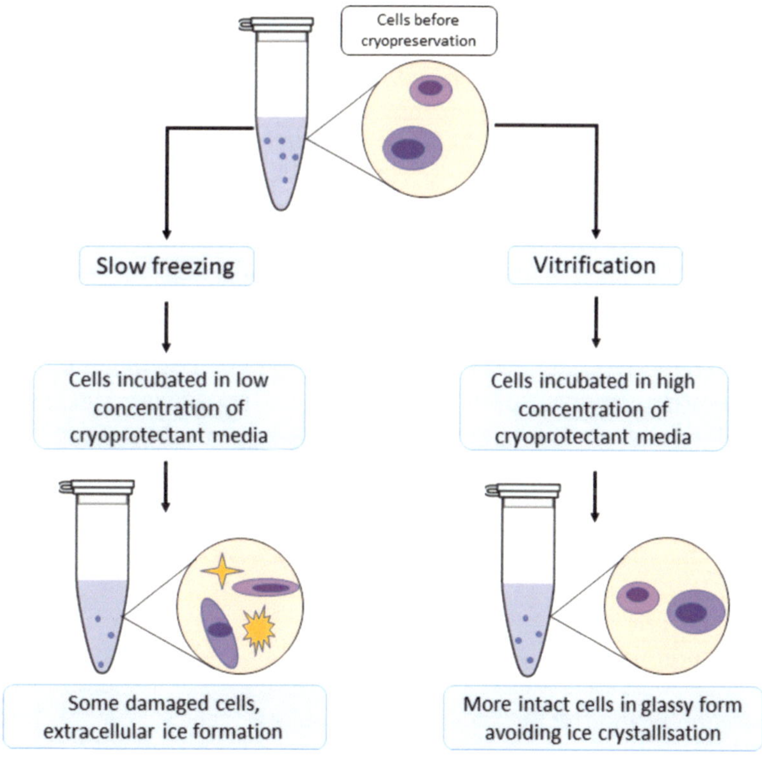

Figure 1.
Comparison of slow freezing and vitrification method.

2.1.1 Slow freezing

Slow freezing involves progressive cooling of sample over a period of 2–4 h either manually or automatically using a semi programmable freezer. This method was developed by Behrman and Sawada [20]. In SF, a phase transition occurs from liquid to solid on temperatures below freezing point. Slow-cooling protocols involve the use of <1.0 M of cryoprotective agents (CPAs), such as glycerol or dimethyl sulphoxide (DMSO), which have minimal toxicity at lower temperatures with the use of a high-cost controlled rate freezer or a benchtop portable freezing container [21].

Main advantage of SF is the reduced risk of contamination during the procedure, without the need of highly skilled professionals. However, SF has many disadvantages. It is time consuming and expensive. There is a high risk of cryoinjury due to the formation of extracellular ice. Although, slow cooling considered a successful method, the success rate is considerably low and might not be suitable for all kinds of cells and tissues [22]. SF is a commonly used method for preservation of animal germplasm for majority of farm animals like sheep, cow, zebrafish etc. [23, 24]. Though with certain drawbacks, nonetheless SF was found more efficient method to cryopreserve ovine embryos in comparison to vitrification [25].

2.1.2 Vitrification or ice-free cryopreservation

To circumvent the process of ice crystallization, cost and duration for cryopreservation in SF, an alternative technique called rapid freezing or vitrification (Vit) was developed. Vit is an ultrarapid cooling method with high cooling rate, which enables putting cells at cryogenic temperatures, forming what is known as ice-front status, while avoiding ice crystallization, hence also termed ice-free cryopreservation. Vitrification is now the recommended protocol for embryos and cells freezing. The process includes cooling the cells or tissue to cryogenic temperatures using liquid nitrogen, after their exposure to high concentrations of CPAs, with subsequent rapid cooling to avoid ice nucleation [11]. The cell suspensions are transformed directly from the aqueous phase to a glass state upon exposure to LN_2. High CPA concentration and higher cooling and warming rates eliminates ice formation [26]. Several factors like viscosity or thickness of the sample, cooling and warming rates, sample volume, CPA type and concentration have considerable effect on the process. Thus, a balance of all the important factors is required for successful vitrification. Vitrification is of two types—equilibrium and nonequilibrium. Formulation of multimolar CPA mixtures and their injection into the cell suspensions is termed as Equilibrium vitrification. Non-equilibrium vitrification includes the use of carriers like plastic straws or cryoloops for obtaining minimum drop volume and carrier-free systems that employs higher freezing rate and lower CPA mix concentration. In comparison to slow freezing, vitrification is more advantageous as is associated with decreased risk of cryoinjury, thereby ensuring sufficiently higher cell survival rate [11]. To note, there is no universal protocol followed for the animal species. However, germ cells from species like mouse ovary [27], fish embryos [28], ovarian sheep [29] and many other species has been cryopreserved by vitrification [24].

2.2 Other emerging cryopreservation techniques

2.2.1 Controlled vitrification by liquidus tracking (LT)

Liquidus tracking (LT) is a slow and controlled vitrification protocol with gradual increase in the cryoprotectant concentration simultaneously with continuous decrease of the temperature at subzero ranges in a specified rate. With LT, recovery and restoration of chondrocytes was achieved successfully from cryopreserved articular cartilage [30] and promising results has been reported in case of ovarian tissue cryopreservation [31]. The principle of LT is the dynamic control of CPA concentrations, throughout the cooling process, in order to maintain the cell just above its freezing point at all times, without the formation of ice [11]. An example, Ovarian Tissue Cryopreservation (OTC) by LT has been reported successful in restoration of ovarian function in sheep model [29].

2.2.2 Laser pulse vitrification

Fowler and Toner proposed the applicability of laser light in cryopreservation process. A successful recovery of rapidly frozen red blood cells by vitrification was achieved without the use of CPAs. Principally, laser targets only the intracellular ice causing it to melt and resolidify into glassy state. After thawing and use of laser light, around 80% of cells treated remained viable [32]. In an attempt, Jin *et al.,* [33] reported full survivals of mouse oocytes after vitrification in 3-fold diluted media and ultra-rapid warming by an IR laser pulse [33].

2.2.3 Isochoric and hyperbaric cryopreservation

New approach in the cryopreservation method is freezing under pressure. Previous method discussed above employed constant-standard pressure (isobaric) conditions near 1 atm of pressure. Isochoric (constant-volume) cooling provides means to significantly lower nonfrozen storage temperatures without any or with only minimum requirements for CPAs, achieving greater metabolic reduction without injury associated with freezing, CPA toxicity, or increased amounts of osmotic solutes. The isochoric cryopreservation is a two-phase equilibrium process, in which ice and liquid exist simultaneously at equilibrium under constant temperature and volume, while hyperbaric cryopreservation the solution is maintained in a single phase as liquid, the survival rate is however low in this method [34]. RBCs were successfully cryopreserved by this method [35]. Despite multiple attempts, scientists have not been able to cryopreserve and restore normal functions of complex bio-samples, such as mammalian tissues and organs.

2.3 Fundamentals of cryopreservation and cryoinjury

Prior to the understanding of the role of cryoprotective agents (CPAs), the impact of subzero temperatures on healthy tissues and basic principles in cryobiology must be knowledged. As known, water is one of the most essential elements present in every cell, tissue, and organ of the living organisms on earth. It constitutes around 80% of tissue mass [36]. On lowering temperature, water undergoes phase transition (liquid to solid) and results in ice crystallization. The formation of intracellular ice cause damage to cellular structure and its function consequently leads to cryoinjury. Freezing can cause two types of harmful effects on cells. The formation of ice crystals damages the cell membrane and thus regain of structurally intact cells on thawing would be difficult. Further, ice formation increases the solute concentration leading to osmotic imbalance and cellular damage. To minimize or to mitigate these effects, two protective actions viz. selection of effective cryoprotectant, and appropriate cooling and thawing rates must be undertaken.

2.3.1 Cryoprotective agents (CPAs)

The discovery of CPA and its role in reducing cryoinjury was a significant step in cryopreservation success. Biological acceptability, cell penetration, low toxicity, are some of the properties, a CPA should possess. As mentioned in previous section, best survival rate of cells and tissues depends on the optimization of factors like cooling rate, warming rate, sample volume. CPA concentration is a major factor influencing the success of the cryopreservation [37]. Based on their penetrating capabilities through cell membrane, CPAs are classified into two categories- membrane permeable/permeating and membrane impermeable/non permeating.

2.3.1.1 Permeating CPAs

Permeating CPAs are smaller sized (typically less than 100 daltons), and amphiphilic in nature [38]. Owing to these properties these molecules can penetrate through the cell membrane easily, tend to equilibrate within the cytoplasm and replace the intracellular water in order to avoid excessive dehydration. Henceforth, they protect the cell from intracellular ice formation (IIF) and salt accumulation. Examples of CPAs in this category are: glycerol (the first agent discovered), dimethyl sulfoxide (DMSO), ethylene glycol (EG), and propanediol (propylene glycol) [39]. The protective role of CPAs is due to hydrogen bonding with water molecules, and lowering the freezing point of water. As a result, less water molecules are available to interact with themselves to form critical nucleation sites required for crystal formation [40]. To minimize toxicity, vitrification mixtures are often added in a stepwise fashion at temperatures near 0°C. Addition of permeating agents prevent the formation of ice and permit cell storage at supercool temperatures. Besides vitrifying, few CPAs, example DMSO have properties to enhance the cell permeability in a dose dependent manner. DMSO of about 5% is reported to increase permeability by reducing the thickness of the cell membrane. DMSO at 10% concentration is more effective and commonly used as it induces water pore formation in biological membranes. Intracellular water can thus easily be replaced by CPAs to promote vitrification. At higher, toxic concentrations (40%) however, lipid bilayers begin to disintegrate [41]. Thus selection of appropriate CPA concentration is vital for maintaining structural integrity and viability after freezing.

2.3.1.2 Non-permeating agents

Unlike permeating, non-permeating agents are covalently linked dimers, trimers or polymers with a larger size. They cannot pass through the cell membrane and exert their protective effect extracellularly. Most commonly used non-penetrating CPAs are polyethylene glycol (PEG), polyvinylpyrrolidone (PVP), raffinose, sucrose, and trehalose [42, 43]. The mode of action of non-permeating agents is similar to permeating agents by controlling osmolarity but works extracellularly and at a lower degree.

List of different permeating and non-permeating cryoprotective agents used in gamete and embryo cryopreservation in different species is listed in **Table 1**. Besides these commonly used CPAs, protein like sericin from silkworm and small antifreeze proteins derived from marine teleosts or fishes have also garnered attention as CPAs in cryobiology [11].

2.3.2 Cooling and thawing rates

Choice of appropriate cooling and thawing rates is another vital step for a successful cryopreservation. Mazur [44] has previously demonstrated the significant

Cryoprotectant	Cell type	Species
Permeating CPAs		
Dimethyl sulphoxide	Oocyte, ovarian tissue, embryo, sperm	Mouse, cow, human
Amides and imides		
Formamide	Sperm	Mouse, horse, dog, goose
Acetamide	Oocyte, embryo	Mouse, rat, rabbit, pig
Propionamide	Sperm	Rabbit
Lactamide	Sperm	Rabbit
Butyramide	Sperm	Rabbit
Malonamide	Sperm	Rabbit
Alcohols		
Methanol	Sperm	Horse
Ethylene glycol	Oocyte, embryo, sperm	Mouse, cow, human
Glycerol	Oocyte, embryo, sperm	Mouse, cow, human
Xylitol	Embryo	Rat
Arabitol	Embryo	Rat
Erythritol	Embryo	Rat
Non-permeating CPAs		
Monosaccharides		
Glucose	Sperm	Cat
Galactose	Sperm	Horse
Disaccharides		
Sucrose	Oocyte, embryo, sperm	Mouse, cow, human
Trehalose	Oocyte, embryo, sperm	Mouse, cow, human
Lactose	Sperm	Mouse, cow
Maltose	Sperm	Mouse, rabbit
Polysaccharides		
Raffinose	Oocyte, sperm	Mouse, horse
Dextran	Embryo	Mouse, cat
Macromolecules		
Polyethylene glycol	Oocyte, embryo	Mouse, human, cow
Ficoll	Oocyte, embryo	Mouse, human, cow
Polyvinyl alcohol	Oocyte, embryo	Mouse, sheep, cow
Hyaluronan	Embryo, sperm	Mouse, sheep, cow

Table 1.
List of different permeating and non-permeating cryoprotectants used in assisted reproductive technologies.

correlation between cooling rates and survivability of various cells [44]. The cooling rate was directly proportional to intracellular ice formation and inversely proportional to survivability among various cells. **Figure 2** illustrates the survival of several cell types after cryopreservation in relation to the cooling rate. All cells exhibited a characteristic inverted U-shaped curve indicating that the survival rate increases with an increase in cooling rate upto a point after which the survival gradually decreases. Larger cells dehydrate slowly compared to smaller cells.

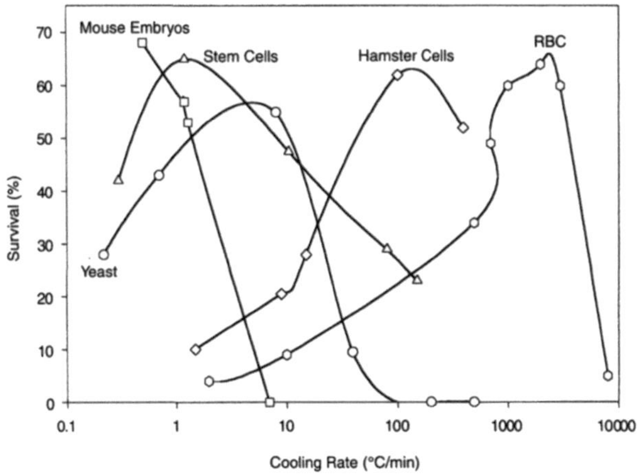

Figure 2.
An inverted U shape curve demonstrating relationship between cooling rate and cell survival (adapted from article by Cipri et al., [45]; figure created originally by Critser and Mobraaten [46] in ILAR Journal).

Hence, in the light of this, rates of cooling and thawing should be adjusted. With an exception to larger cells, a cooling rate of approximately 1°C/min is often recommended. Controlled rate freezers that modulate chamber temperatures are used for this purpose. Following cryopreservation, cells are stored for future thawing and appropriate use.

Unlike cooling rate, thawing rate has been given inadequate attention. Nevertheless, it is advisable to warm or thaw cells rapidly to prevent recrystallization of ice [47]. This can be explained thermodynamically as vitrified state is quasi-stable and can change into a lower energy crystal structure on thawing. Currently, to achieve maximum viability it is suggested to transport cryovials on vapor LN_2 and warmed rapidly in a 37°C water bath for 90–120 s [39]. Decrease in viability after post thaw recovery is inevitable, no matter how well the cells were stored and thawed. Care should be taken to remove dead cells timely using density media like Ficol or other methods from the recovered cells to increase the viability of recovered cell prior to use. Density gradients can be utilized to increase viable cell density although this method often involves exposing cells to additional, potentially-harmful centrifugation. Current strategies for identifying cells that remain viable after preservation utilize organic fluorophores, and dyes.

2.3.3 Recent advances in post-cryo cell recovery

The complete cryopreservation procedure is associated with great efforts to maintain cell viability and function. However, some of the cryopreserved cells often demonstrate decreased viability following thawing. Therefore, it is essential to remove dead cells and increase the viability of live cells. Currently, organic fluorophores are employed for viable cells selection. Viable cell selection using this method is commonly practiced in ART in animal reproduction example semen/spermatozoa of bull, buffalo, rabbit, alpaca, stallion/horse and many other species [48–50]. The earlier practice to achieve this, involved the use of density gradient. Although a little expensive, advanced technologies like magnetic affinity cell separation (MACS), and fluorescence activated cell sorting (FACS) appeared beneficial [51, 52]. Cryosurvival and recovery of sperm in species like stallion, bovine etc. using FACS has been investigated [53, 54]. New strategies like the introduction of nano-science had revolutionized the field of cryobiology and forwarded it to more higher stages as nanoparticle mediated cell sorting is non-destructive and more beneficial than FACS [55].

2.4 Sperm cryopreservation

After the successful live birth with 21 years cryopreserved sperm sample, sperm viability in cryopreserved sample was evident [56]. Currently, sperm cryopreservation as a method of fertility preservation had gained tremendous significance and applicability in human and animals [57, 58]. Sperm survival rate was found to increase with glycerol as CPAs and increasing concentration of glycerol added at constant cooling rate for long term storages [59]. Sperm freezing can be done by one of the main two established methods; slow freezing or ultra-rapid freezing [60, 61]. Owing to the deleterious effects of SF on sperm physiochemical activity and motility, the rapid freezing approach was suggested to be the potential solution to preserve cells without allowing the nucleation of ice crystals. Currently, ultra-rapid freezing adopted ensures both intra and extracellular vitrification thus improving sperm survival [62].

2.5 Female fertility cryopreservation

Like sperm, female germ cells can be frozen by cryopreservation. Fertility preservation in female cancer survivors is a major concern in oncology and assisted conception. Fertility in females can be preserved by freezing egg either embryo, oocytes and oocytes within ovarian tissue. The results with oocyte cryopreservation were unsatisfactory. Ovarian tissue cryopreservation on the other side have attracted the interest of the medical and scientific communities. Cryopreservation protocols for oocyte and ovarian tissue are discussed briefly below.

2.5.1 Oocyte cryopreservation

For oocyte cryopreservation, currently acknowledged methods are SF using equilibrium freezing and Vit/non-equilibrium protocols. In cryopreserving oocytes by SF method, the protocol involves gradual cooling of the specimen to lower temperature (−150°C)with controlled slow cooling rates in presence of low concentration of DMSO (1.5 M) plus non-permeating sugars like sucrose or trehalose at 0.3 M concentration. Specimens are stored at −196°C in LN_2. The survival rate achieved by SF remains relatively unsatisfactory. Evidence from earlier investigations indicates survival rates plateau around 70–80% with this method [63]. Hence, the recommended method for cryopreservation of oocytes is vitrification. Initially, oocytes are equilibrated with a solution containing PEG and DMSO at 7.5% v/v for 5–15 min. Prior to storing the cells in $LN_{2,}$ the cells are exposed to vitrification media (PEG and DMSO-15% v/v, -and sucrose (0.5 M) for a minute and stored. For thawing, CPAs are gradually removed to avoid ice crystal formation. Later, the cells are revived following incubation in culture medium [64]. Intriguing results were obtained in systematic analysis conducted by Arav and Natan. Vitrified oocytes were reported to have higher oocyte survival and fertility rates compared to slow-cooled oocytes. Furthermore, no differences were observed in pregnancy rate, formation of top quality embryo and fertilization between vitrified and fresh oocytes thus strongly signifying vitrification as the superior procedure for the oocytes cryopreservation [65]. During the past few years, a significant progress has been made in the cryopreservation of oocytes of different species using new vitrification methods. High rates of survival and development after solid-surface vitrification have been reported for *in vitro* matured oocytes from cows [66] and goats [67].

2.5.2 Ovarian tissue cryopreservation (OTC)

Parrot's study, which resulted in first mice offspring was developed following ovarian tissue cryopreservation (at −79°C) and isografting was a breakthrough in the early 1960s [68, 69]. Later, with further discoveries of CPAs like DMSO, propanediol, and ethylene glycol, cryopreservation methods gradually improved. Like oocyte, ovarian tissues are also preserved by SF and Vit method, specially ovarian cortical tissue from mammals [70]. After the advent of Vit for oocyte cryopreservation and in comparison, to SF, ovarian tissue vitrification is now considered promising for ovarian cortical tissue cryopreservation [71, 72]. Ovarian tissue cryopreservation is successfully accomplished in many animal species like mice [68], ewe [73], cow including human [74] and other species.

Table 2 shows the comparison between SF and vitrification protocol for ovarian tissue cryopreservation. Vitrification was suggested to have several advantages over the conventional SF in characteristics reviewed. However, SF remains the standard clinical method until further reports show improved success rate for Vit to be applied in human clinical use over SF. Currently, the main purpose for OTC is fertility preservation known so far, especially in young women diagnosed with cancer or some genetic disease that destroys ovarian reserve. Access to immature oocytes from antral follicles and restoration of organ function have evoked new perspectives in utility of OTC for social reasons besides medical use.

2.6 Applications of fertility preservation

Germ cell cryopreservation certainly had emerged as effective method for long term fertility preservation in the field of reproductive medicine in both human and animals. Germ cells cryopreservation would be applicable in restoring fertility in animals and humans and preservation of endangered animal species. Over a decade there has been increase in production of GEA models from cryopreserved animal genetic resources for disease investigations. The practice of genetic engineering has increased the number of mouse and rat lines to tenfold the actual number. Currently, sperm cryopreservation is a fundamental technique in assisted reproduction technology (ART) like artificial insemination (AI), *in vitro* fertilization (IVF), or intra-cytoplasmic sperm injection (ICSI). AI has relatively been the

Features	Slow freezing	Vitrification
Period of freezing process	Slow usually 3–4 h	Fast <10 min
Direct contact with liquid N_2	No	Yes
Ice formation	Yes	No
CPA concentration	Low	High
Cooling rates (°C/min)	0.15–0.3	15,000-30,000
Cost	Expensive	Inexpensive
Special equipment	Yes	No
Techical expertise	Simple	Risky
Sample volume	100–200 μl	1–2 μl
Mechanical damage	More	Less
Chemical change	Less	More

Table 2.
Comparison between SF and vitrification protocol.

most important practice contributing to the advancement of animal production. Many advantages of AI are enhanced when semen is cryopreserved and stored for extended periods. Several species such as mice, rabbits, pigs, goats, cows, and sheep has been successful reproduced adopting egg and ovarian tissue cryopreservation [66–68, 73, 74]. After successful animal experimentation using cryopreserved germ cells, utility of fertility preservation is gradually extended to human. At present, cryopreservation of human oocytes and sperm has been carried out for medical or social reasons. Infertility in couples is solved to a considerable extend. Cancer patients who are at risk of fertility loss either due to radio or chemotherapy and women who wish to conceive in later ages can benefit from the cryopreservation technology and fertility restoration.

2.7 Frontiers in cryopreservation in animal models

The birth of mouse from 50 years old cryopreserved embryo had revolutionized the field of animal reproduction. Later, to achieve greater reproductive outcomes, cryopreservation protocols have been continuously refined over the years. In the given section, few of the important animal models has been discussed.

2.7.1 Mouse

Mouse is one of the most commonly used animal model for research. In early 1990s, the first attempt in freezing mouse was a grand success. Later, cryopreservation of mouse spermatozoa resulted in production of a large number of mouse inbred and hybrid strains [75, 76]. There is a growing demand for the production of genetically modified mouse strains since mouse has become the most profound model system to investigate the genetics and pathogenetics of human diseases. Moreover, knock-out projects started in Europe and USA over a decade. As life maintenance of the growing number of mice is difficult and uneconomical in animal laboratories, germplasm cryopreservation provides a valuable means of maintaining transgenic mouse strains used in biomedical research [77]. Moreover, animal gametes- sperm, eggs, and embryos preservation are now successfully preserved and maintained in cryobanks owing to the cryopreservation technique. Historically, embryo cryopreservation served as the gold standard for maintaining transgenic mice strains with single, multiple mutations, or complex genetic background [78, 79]. However, it is often more expensive due to costly and time-consuming superovulation procedures and subsequent cryopreservation. However, mouse sperm cryopreservation for long-term storage is simple and inexpensive, and it only requires few donor animals for protecting those commonly used inbred strains (e.g., C57BL/6, FVB, and 129/Sv) with single mutations [80]. Thus, sperm cryopreservation provides an efficient management of these genetic resources by reducing maintenance space and cost and by safeguarding them against, for example, disease, breeding failure, and genetic drift.

Although current sperm cryopreservation protocols showed relatively high success, there has been variation in the sperm motility and IVF outcomes post thawing. Many researchers investigated and worked to develop the gold standard protocol for mouse sperm freezing. Nakagata's protocol became the most widely used by many research laboratories and clinical and scientific facilities around the world [80]. The initial freezing solution simply contains 18% dehydrated skim milk and 3% raffinose in water, and cooling is achieved in LN_2 vapor phase for 5 min followed by plunging the samples into LN_2 at −196°C. Since the introduction of this initial protocol, there have been few changes in an effort to improve post-thaw fertilization potential of mouse sperm.

For sperm cryopreservation in mouse, epididymal sperm collected from the cauda epididymis of matured male mice are suspended in sugar-based cryoprotectant, which is loaded in freezing straws and preserved at −196°C [81]. The cryopreserved sperm can be used for efficient fertilization using improved IVF systems featuring methyl-β-cyclodextrin (MBCD) and reduced glutathione (GSH) [82, 83].

Several scientist across the globe had investigated the role of varied antioxidants like monothioglycerol [84], methyl-β-cyclodextrin (MBCD) [82] and the latest refinement was the introduction of reduced glutathione (GSH) to protect spermatozoa against oxidative stress during IVF treatment [83] and increase the fertilization rates. C57BL/6 is a major inbred strain used for the production of transgenic mice, and also as a back-cross for targeted mutant mice [85]. Therefore, it is necessary to establish cryopreservation method for C57BL/6 mouse spermatozoa that could maintain a high fertilizing ability after thawing. Recently, a preincubation medium containing methyl-beta-cyclodextrin used demonstrated increased fertility [82]. Further, it is augmented that sperm freezed in a cryomedia containing 18% raffinose and 3% skim milk, increased fertilization rate [86]. Sperm banking can be used for mice, but in some instances it is also important to bank the female genome [87].

Further, ovarian gamete cryopreservation, as harvested oocytes or contained within the primordial follicles in the cortical patch tissue, can be used for long-term storage of the female germline [88]. Cryopreservation protocol for ovarian tissue based on the slow-cooling procedures was initially developed and used for mouse eggs and embryos in 1970s [10]. This process requires a biological controlled rate freezer or equivalent equipment. It is now documented that in the mouse both fresh and frozen thawed grafts have the potential to restore long-term fertility (i.e., for 1 year) to the graft recipient [89]. A more recent advancement in duration of graft function is in case of ovarian tissue transplanted with transplanted graft functional for 5 years, or more persisting for more than 9 years [90]. Slow freezing is overshadowed with the development of more advanced closed vitrification system, proven more beneficial in ovarian cryopreservation.

2.7.2 Rat

Animal model that is also commonly involved in the scientific work is rat, mainly in basic biology, physiology, brain science, and medicine. Over 40 year ago, the success of IVF was reported in rats. Characteristics, such as a short gestation and a relatively short life span, docile behavior, and ready availability of animals with well-defined health and genetic backgrounds make rat an ideal experimental model. The rat is a standard species for toxicological, teratological, and carcinogenesis testing by the pharmaceutical industry and governmental regulatory agencies [91, 92]. Rats are still continued to be used for nutritional, neurological and endocrinology studies. Cryopreservation has not been performed in rats as often as it has in mice, but the technique is becoming more widespread, for the same reasons that it is used widely in mice. Cryopreservation can be an efficient method of maintaining the potential of raising live mice of the thousands of genetically modified genotypes currently available [93, 94]. Sperm cryopreservation has been similarly successful in rats. Offspring are obtained from thawed sperm using intrauterine insemination or via IVF. Seita *et al.,* [95] established for the first time a successful IVF system using cryopreserved rat sperm [95]. A generalized protocol for IVF in rats (and similar breeds) from cryopreserved sperm and oocytes is illustrated in **Figure 3**.

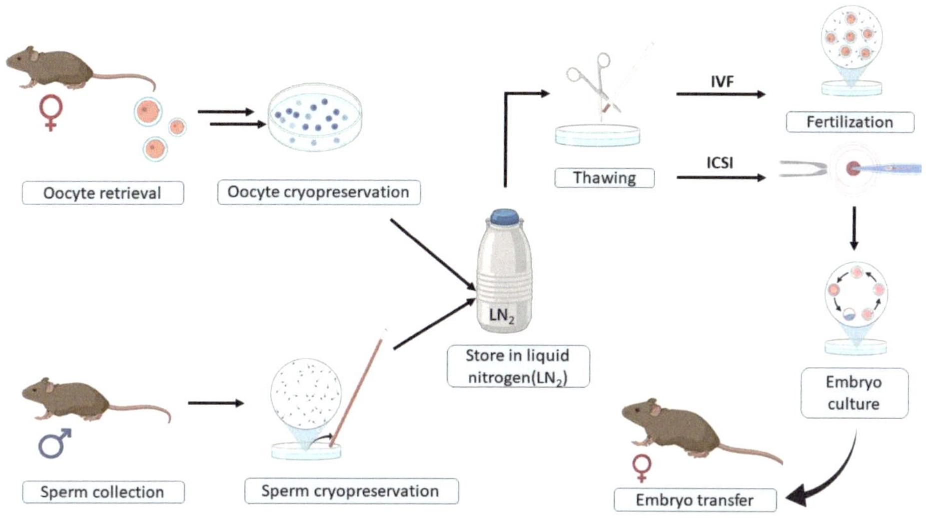

Figure 3.
A schematic representation of Assisted Reproduction techniques and In vitro fertilization; IVF in rats from cryopreserved sperm and oocytes.

2.7.3 Rabbit

AI in rabbit using cooled semen stored for short time is a commonest ART practice in country like Europe, where rabbit rearing is common [96, 97]. High fertility rates and proliferation are obtained in this process. However, with limiting factors like low productivity and issues with cryopreserved rabbit sperm, AI is used for experimental or genetic resource bank purposes. Although cryopreserved sperm is not used for commercial purposes at the present, there is a need for reliable methods of rabbit sperm resource banking, especially as this species is a valuable animal in therapeutics(for production of vaccines, antibodies, hormones) [98, 99]. The extent of cryoinjury varies with the species due to differences in the gamete plasma membrane composition among them and cell size as well [100]. Hence, there is no universal protocol followed for all the animal species. It is obligatory to standardize cryopreservation protocol, extender and CPAs levels and composition, for each single species or even breeds.

Researchers observed significantly varying outcomes with the type of CPAs used in cryopreserving rabbit sperm. The use of CPAs like DMSO or ethylene glycol is recommended to enhance the storage of rabbit sperm [101]. Few other studies observed ethylene glycol to be less toxic than glycerol in cryofreezing rabbit sperm. Despite lower toxicity, ethylene glycol failed to provide protection to sperm cells when compared to other DMSO. Importantly, CPAs concentration need to be optimized for sufficient protection. To note, addition of CPAs at 5°C instead of 37°C was reported effective in cryopreserving rabbit sperm as permeability increases with increased temperatures [102]. Data on the CPA type and concentration for optimal cryopreservation is contradictory. However, it is well documented that a balance between all the involved parameters is a key for successful freezing. Sperm quality varies with the type of extenders used. The extenders used for rabbit sperm cryopreservation include a mixture of permeable and non-permeable CPAs. The use of two permeable CPAs is a common practice [103]. Sperm frozen in acetamide extender was found to be of superior quality than sperm DMSO and glycerol or other mixtured extenders [104]. The mixture of CPAs with egg yolk has differential effect on the sperm quality. Sperm frozen in extender containing egg yolk and

DMSO demonstrated better sperm quality than for sperm frozen in the extender that contained high DMSO and lacking egg yolk [102]. Hence, it is obvious that a balance of all the essential factors is a key for successful freezing.

Recently, Domingo *et al.*, [105] studied comparison of different semen extenders and cryoprotectant agents to enhance cryopreservation of rabbit spermatozoa and found that the addition of dimethylformamide (DMF) to INRA 96® exerts a protective effect on the membrane of spermatozoa improving seminal quality [105]. Although many efforts have been made to optimize cryopreservation extenders and protocols for rabbit sperm, many questions remain unanswered. In addition to cryopreserving sperm, rodents serve as a model for ovarian tissues cryopreservation and transplantation procedures as for current and future application and clinical use in the human. The production of GE rodent models for disease research increased over a decade using gene editing technologies (like CRISPR/Cas9).

2.7.4 Cryopreservation from rodents to larger animals

From rodents to larger animals, cryopreservation is proved to be beneficial in fertility preservation, transplantation and breeding livestock. Advances in cryopreservation pioneered with transplantation of cryopreserved mouse primordial follicles in 1993 by J. Carroll and R. Gosden [106] followed by recognition of sheep as a larger model to study ovarian function. Gosden, in collaboration with D. Baird [107], developed techniques with vascular anastomosis that formed the basis of transplantation of larger organs such as kidney in human. The development of ART techniques has gained significance in the production of commercially important farm animal breeds and a few exotic or endangered species. Livestock industry especially cattle had benefitted to a major extent from the application of cryopreserved semen or embryos over the past decades. This was also the case in experimenting gamete cryopreservation. Larger animal like sheep is greatly emphasized to study human diseases particularly respiratory diseases and lung cancer, since the anatomy and physiology of the sheep respiratory system is more similar to that of humans than rodents. Sheep has been proposed as a good model for vaccines, asthma pathogenesis and inhalation treatments [108]. The gradual decline of genetic diversity in domestic breeds imposes a major threat to livestock, hence, international community strives harder to conserve the livestock genetics. Semen from most of the mammalian species has been successfully frozen [109].

With regards to temperature tolerance, sperm from different species exhibits varied responses. Bovine sperm shows higher tolerance to low temperatures, while porcine and ovine sperm are more sensitive and at risk of cold shock when exposed to temperature between 5° and 22°C leading to rapid loss of vitality. Animal sperm is highly vulnerable to oxidative damage owing to the loss of antioxidase enzyme and increased fatty acid oxidation on freezing [110]. The stability and viability of sperm is enhanced by adding semen extenders during freezing. The first semen extender for bovine sperm preservation used was egg yolk-sodium citrate diluent (EYC) and was gradually replaced with tris-buffered egg yolk (TRISEY)or tris-fructose yolk-glycerol [111]. Most of the industries use tris and citrate as active components in bovine sperm extenders. Addition of compounds like vitamin E to semen extenders is found to increase the structural integrity of acrosome, and thereby preventing sperm from oxidative damage via its antioxidant properties [24].

2.7.5 Vitrification in larger animals

Vitrification of embryos was invented in 1985 [112]. Later vitrification emerged as one of the powerful methods for cryopreserving embryo from farm animals including cattle, goat, sheep and pig [28, 29, 113, 114]. The birth of the first calf was achieved from frozen/thawed oocyte was reported in 1992 [115]. Vitrification showed high success in bovine oocyte cryopreservation in 1998 [116]. Applicability of macromolecules with lesser toxicity as CPAs, the use of cryotop and solid surface virtification emerged gradually over time to overcome cryoinjury. Treatment with docetaxel improved cryopreservation of bovine oocyte as its protective against cytoskeleton injury thus can potentially enhance survival rate of post thawed oocytes [117]. High rates of survival and development after solid-surface vitrification have been reported for *in vitro* matured oocytes from cows [66] and goats [67]. SSV uses a metal surface, precooled to −180°C by partial immersion into liquid nitrogen, to cool microdrops of vitrification solution containing the embryos or oocytes.

OTC has been reported successful in restoration of ovarian function in sheep model. Complete restoration of acute ovarian function and high rates of natural fertility with multiple live births, were obtained following whole ovary cryopreservation and autotransplantation of adult sheep ovaries [29]. Although ovarian tissue cryopreservation has developed from experiments in sheep in early 1990s, it is now becoming recognized as relatively successful procedure for OTC in human, particularly to preserve the fertility of cancer patients to avoid gonadotoxic damage resulting from the therapy [118]. While there has been considerable success with cryopreservation of oocytes, embryos and semen in farm animals, this technology still requires refinement and further optimizing studies.

3. Conclusion

Based on the utility and need of the animals for varied purposes, fertility preservation is a prerequisite to be practiced in animal husbandry and animal house unit for production of animal models. Understanding of the fundamentals of cryopreservation allows the development of more efficient procedures for cryopreservation of germ cells and further expand their clinical applications and utility in livestock, which can also be transferred to human application. Although, the field of cryobiology has advanced over the years, further research remains required to optimize cryoprotectant concentration, cooling and thawing rates to aim high success in animal reproduction.

Author details

Feda S. Aljaser[1,2]

1 Department of Clinical Laboratory Sciences, College of Applied Medical Sciences, King Saud University, Riyadh, Saudi Arabia

2 Chair of Medical and Molecular Genetics Research, Department of Clinical Laboratory Sciences, College of Applied Medical Sciences, King Saud University, Riyadh, Saudi Arabia

*Address all correspondence to: faljaser@ksu.edu.sa

References

[1] Baumans V. Use of animals in experimental research: An ethical dilemma? Gene Therapy. 2004;**11**:S64-S66. DOI: 10.1038/sj.gt.3302371

[2] Hickman DL, Johnson J, Vemulapalli TH, Crisler JR, Shepherd R. Commonly used animal models. In: Principles of Animal Research for Graduate and Undergraduate Students. United States: Elsevier; 2017. pp. 117-175. DOI: 10.1016/B978-0-12-802151-4.00007-4

[3] Ancel C, Bentsen AH, Sebert ME, Tena-Sempere M, Mikklesen JD, Simonneaux V. Stimulatory effect of RFRP-3 on the gonadotrophic axis in the male Syrian hamster: The exception proves the rule. Endocrinology. 2012;**153**:1352-1363. Available from: https://books.google.com.sa/books?id=fvaZBQAAQBAJ&source=gbs_navlinks_s

[4] Le Bras A. A resource for selecting animal models of heart disease. Laboratory Animals. 2019;**48**:332. DOI: 10.1038/s41684-019-0425-4

[5] Morrison JL, Botting KJ, Darby JRT, David AL, Dyson RM, Gatford KL, et al. Guinea pig models for translation of the developmental origins of health and disease hypothesis into the clinic. The Journal of Physiology. 2018;**596**(23): 5535-5569. DOI: 10.1113/JP274948

[6] Padilla-Carlin DJ, McMurray DN, Hickey AJ. The Guinea pig as a model of infectious diseases. Comparative Medicine. 2008;**58**:324-340. Available from: https://www.ncbi.nlm.nih.gov/pmc/articles/PMC2706043/

[7] Bahr A, Wolf E. Domestic animal models for biomedical research. Reproduction in Domestic Animals. 2012;**47**:59-71. DOI: 10.1111/j.1439-0531.2012.02056.x

[8] Agca Y. Genome resource banking of biomedically important laboratory animals. Theriogenology. 2012;**78**(8): 1653-1665. DOI: 10.1111/j.1439-0531.2012.02056.x

[9] Ugur MR, Saber Abdelrahman A, Evans HC, Gilmore AA, Hitit M, Arifiantini RI, et al. Advances in cryopreservation of bull sperm. Frontiers in Veterinary Science. 2019;**6**:268. DOI: 10.1016/j.theriogenology.2012.08.012

[10] Gao D, Critser J. Mechanisms of cryoinjury in living cells. ILAR Journal. 2000;**41**:187-196. DOI: 10.1093/ilar.41.4.187

[11] Jang TH, Park SC, Yang JH, Kim JY, Seok JH, Park US, et al. Cryopreservation and its clinical applications. Integrative Medicine Research. 2017;**6**(1):12-18. DOI: 10.1016/j.imr.2016.12.001

[12] Priyanka V, Kumar R, Dhaliwal I, Kaushik P. Germplasm conservation: instrumental in agricultural biodiversity—A review. Sustainability. 2021;**13**(12):6743. DOI: 10.3390/su13126743

[13] Leroy G, Boettcher P, Besbes B, Danchin-Burge C, Baumung R, Hiemstra SJ. Cryoconservation of animal genetic resources in Europe and two African countries: A gap analysis. Diversity. 2019;**11**(12):240. DOI: 10.3390/d11120240

[14] Tao Y, Sanger E, Saewu ALeveille MC. Human sperm vitrification: The state of the art. Reproductive Biology and Endocrinology. 2020;**18**:17. DOI: 10.1186/s12958-020-00580-5

[15] Walters EM, Benson JD, Woods EJ, Crister JK. The history of sperm cryopreservation. In: Pacey AA, Tomlinson MJ, editors. Sperm Banking Theory and Practice. London (UK): Cambridge University Press; 2009. pp. 1-17. Avaialble from: https://assets.

cambridge.org/97805216/11282/excerpt/9780521611282_excerpt.pdf

[16] Spallanzani L. Opuscoli di fisca animale, e vegatabile. Opuscolo II. Osservazioni, e sperienze intorno ai vermicelli spermatiei dell'suomo edegli animali, Modena. Italy: Presso la Societa Tipagrafica; 1776. Avaialble from: https://books.google.com.cu/books?id=069eAAAAcAAJ

[17] Polge C, Smith AU, Parkes A. Revival of spermatozoa after vitrificationand dehydration at low temperatures. Nature. 1949;**164**:666. DOI: 10.1038/164666a0

[18] Bucak MN, Tekin N. Kryoprotektanlar ve gamet hücrelerinin dondurulmasında kryoprotektif etki. Ankara Üniversitesi Veteriner Fakültesi Dergisi. 2007;**54**:67-72. DOI: 10.1501/Vetfak_0000000255

[19] Whittingham D, Leibo S, Mazur P. Survival of mouse embryos frozen to −196 and −269°C. Science. 1972;**178**:411-414. Available from: https://pubmed.ncbi.nlm.nih.gov/5077328/

[20] Behrman SJ, Sawada Y. Heterologous and homologous inseminations with human semen frozen and stored in a liquid-nitrogen refrigerator. Fertility and Sterility. 1966;**17**(4):457-466. DOI: 10.1016/s0015-0282(16)36003-4

[21] Thompson M, Nemits M, Ehrhardt R. Rate-controlled cryopreservation and thawing of mammalian cells. Protocol Exchange. 2011. DOI: 10.1038/protex.2011.224

[22] Wolkers WF, Oldenhof H. Principles underlying cryopreservation and freeze-drying of cells and tissues. Methods in Molecular Biology. 2021;**2180**:3-25. DOI: 10.1007/978-1-0716-0783-1_1

[23] Marques LS, Fossati AAN, Rodrigues RB, et al. Slow freezing versus vitrification for the cryopreservation of zebrafish (*Danio rerio*) ovarian tissue. Scientific Reports. 2019;**9**:15353. DOI: 10.1038/s41598-019-51696-7

[24] Huang Z, Gao L, Hou Y, Zhu S, Fu X. Cryopreservation of farm animal gametes and embryos: Recent updates and progress. Frontiers of Agricultural Science and Engineering. 2019;**6**(1):42-53. DOI: 10.15302/J-FASE-2018231

[25] Varago FC, Moutacas VS, Carvalho BC, Serapião RV, Vieira F, Chiarini Garcia H, et al. Comparison of conventional freezing and vitrification with dimethylformamide and ethylene glycol for cryopreservation of ovine embryos. Reproduction in Domestic Animals. 2014;**49**:839-844. DOI: 10.1111/rda.12376

[26] Day JG, Stacey GN. Gefriertrocknen. Cryopreservation and freeze-drying protocols. FEBS Letters. 2007;**377**(2):281-282. Available from: https://link.springer.com/book/10.1007/978-1-59745-362-2

[27] Okamoto N, Nakajima M, Sugishita Y, Suzuki N. Effect of mouse ovarian tissue cryopreservation by vitrification with Rapid-i closed system. Journal of Assisted Reproduction and Genetics. 2018;**35**(4):607-613. DOI: 10.1007/s10815-018-1121-9

[28] Keivanloo S, Sudagar M. Cryopreservation of Persian sturgeon (Acipenserpersicus) embryos by DMSO-based vitrificant solutions. Theriogenology. 2018;**85**:1013-1018. DOI: 10.1016/j.theriogenology.2015.11.012

[29] Campbell BK, Hernandez-Medrano J, Onions V, Pincott-Allen C, Aljaser F, Fisher J, et al. Restoration of ovarian function and natural fertility following the cryopreservation and autotransplantation of whole adult sheep ovaries. Human Reproduction.

2014;**29**(8):1749-1763. DOI: 10.1093/humrep/deu144

[30] Pegg D, Wang L, Vauhan D. Cryopreservation of articular cartilage. Part3: The liquidus-tracking method. Cryobiology. 2006;**52**:360-368. DOI: 10.1016/j.cryobiol.2006.01.004

[31] Aljaser F. Gametes cryopreservation using novel ultra-rapid freezing protocols [thesis]. University of Nottingham; 2015

[32] Fowler AJ, Toner M. Prevention of hemolysis in rapidly frozen erythrocytes by using laser pulse. Annals of the New York Academy of Sciences. 1998;**858**:245-252. DOI: 10.1111/j.1749-6632.1998.tb10158.x

[33] Jin B, Kleinhans FW, Mazur P. Survivals of mouse oocytes approach 100% after vitrification in 3-fold diluted media and ultra-rapid warming by an IR laser pulse. Cryobiology. 2014;**68**(3):419-430. DOI: 10.1016/j.cryobiol.2014.03.005

[34] Taylor MJ, Weegman BP, Baicu SC, Giwa SE. New approaches to cryopreservation of cells, tissues, and organs. Transfusion Medicine and Hemotherapy. 2019;**46**:197-215. DOI: 10.1159/000499453

[35] Greer N. Freezing under pressure: A new method for cryopreservation. Cryobiology. 2015;**70**(1):66-70. DOI: 10.1016/j.cryobiol.2014.12.005

[36] Pegg DE. Principles of cryopreservation. In: Day JG, Stacey GN, editors. Cryopreservation and Freeze-Drying Protocols. Methods in Molecular Biology. Totowa (NJ): Humana Press; 2007. pp. 39-75. Available from: https://link.springer.com/book/10.1007/978-1-59745-362-2

[37] Capodanno F, Daolio J, De Feo G, et al. A monocentric analysis of the efficacy of extracellular cryoprotectants in unfrozen solutions for cleavage stage embryos. Reproductive Biology and Endocrinology. 2019;**17**:84. DOI: 10.1186/s12958-019-0519-2

[38] Wowk B. How cryoprotectants work. Cryonics. Scottsdale (AZ): Alcor life extension foundation; 2007. Available from: https://www.alcor.org/docs/how-cryoprotectants-work.pdf

[39] Whaley D, Damyar K, Witek RP, Mendoza A, Alexander M, Lakey JR. Cryopreservation: An overview of principles and cell-specific considerations. Cell Transplantation. 2021;**30**:963689721999617. DOI: 10.1177/0963689721999617

[40] Mandumpal JB, Kreck CA, Mancera RL. A molecular mechanism of solvent cryoprotection in aqueous DMSO solutions. Physical Chemistry Chemical Physics. 2010;**13**(9):3839-3842. DOI: 10.1039/C0CP02326D

[41] Gurtovenko AA, Anwar J. Modulating the structure and properties of cell membranes: The molecular mechanism ofdimethyl sulfoxide. The Journal of Physical Chemistry B. 2007;**111**(35):10453-10460. DOI: 10.1021/jp073113e

[42] Bartolac LK, Lowe JL, Koustas G, Grupen CG, Sjöblom C. Effect of different penetrating and non-penetrating cryoprotectants and media temperature on the cryosurvival of vitrified in vitro produced porcine blastocysts. Animal Science Journal. 2018;**89**(9):1230-1239. DOI: 10.1111/asj.12996

[43] Eroglu A. Cryopreservation of mammalian oocytes by using sugars: Intra- and extracellular raffinose with small amounts of dimethylsulfoxide yields high cryosurvival, fertilization, and development rates. Cryobiology. 2010;**60**(Suppl. 3):54-59. DOI: 10.1016/j.cryobiol.2009.07.001

[44] Mazur P. Kinetics of water loss from cells at subzero temperatures and the likelihood of intracellular freezing. The Journal of General Physiology. 1963;**47**(2):347-369. DOI: 10.1085/jgp.47.2.347

[45] Cipri K, Cioccolanti L, Naldi R. Experimental analysis of a pulse tube based new prototype for cells cryopreservation. International Journal of Energy Research. 2020;**44**:5905-5916. DOI: 10.1002/er.5367

[46] Critser JK, Mobraoten LE. Cryopreservation of Murine Spermatozoa. ILAR Journal. 2000;**41**:197-206. DOI: 10.1093/ilar.41.4.197

[47] Baboo J, Kilbride P, Delahaye M, Milne S, Fonseca F, Blanco M, et al. The impact of varying cooling and thawing rates on the quality of cryopreserved human peripheral blood T cells. Scientific Reports. 2019;**9**(1):3417. DOI: 10.1038/s41598-019-39957-x

[48] Boccia L, Di Palo R, De Rosa A, Attanasio L, Mariotti E, Gasparrini B. Evaluation of buffalo semen by Trypan blue/Giemsa staining and related fertility *in vitro*. Italian Journal of Animal Science. 2007;**6**:739-742. DOI: 10.4081/ijas.2007.s2.739

[49] Valdivia M, Bravo Z, Reyes J, Gonzales GF. Rescue and conservation of male adult alpacas (Vicugna pacos) based on spermatogonial stem cell biotechnology using atomized black maca as a supplement of cryopreservation medium. Frontiers in Veterinary Science. 2021;**8**:597964. DOI: 10.3389/fvets.2021.597964

[50] Kútvölgyi G, Stefler J, Kovács A. Viability and acrosome staining of stallion spermatozoa by Chicago sky blue and Giemsa. Biotechnic & Histochemistry. 2006;**81**(4):109-117. DOI: 10.1080/10520290600931007

[51] Feugang JM. Novel agents for sperm purification, sorting, and imaging. Molecular Reproduction and Development. 2017;**84**(9):832-841. DOI: 10.1002/mrd.22831

[52] Okumura N, Kusakabe A, Hirano H, Inoue R, Okazaki Y, Nakano S, et al. Density-gradient centrifugation enables the purification of cultured corneal endothelial cells for cell therapy by eliminating senescent cells. Scientific Reports. 2015;**5**:15005. DOI: 10.1038/srep15005

[53] Castro LS, Hamilton TRS, Mendes CM, et al. Sperm cryodamage occurs after rapid freezing phase: Flow cytometry approach and antioxidant enzymes activity at different stages of cryopreservation. Journal of Animal Science and Biotechnology. 2016;7:17. DOI: 10.1186/s40104-016-0076-x

[54] Pruß D, Yang H, Luo X, Liu D, Hegermann J, Wolkers WF, et al. High-throughput droplet vitrification of stallion sperm using permeating cryoprotective agents. Cryobiology. 2021;101:67-77. DOI: 10.1016/j.cryobiol.2021.05.007

[55] Durfey CL, Swistek SE, Liao SF, Crenshaw MA, Clemente HJ, Thirumalai RVKG, et al. Nanotechnology-based approach for safer enrichment of semen with best spermatozoa. Journal of Animal Science and Biotechnology. 2019;**10**:14. Available from: https://www.cabdirect.org/cabdirect/abstract/20193275155

[56] Horne G, Atkinson AD, Pease EHE, Logue JP, Brison DR, Lieberman BA. Live birth with sperm cryopreserved for 21 years prior to cancer treatment: Case report. Human Reproduction. 2004;**19**(6):1448-1449. DOI: 10.1093/humrep/deh249

[57] Yeste M. Sperm cryopreservation update: Cryodamage, markers, and factors affecting the sperm freezability

in pigs. Theriogenology. 2016;**85**(1):47-64. DOI: 10.1016/j.theriogenology

[58] Loren AW, Mangu PB, Beck LN, Brennan L, Magdalinski AJ, Partridge AH, et al. Fertility preservation for patients with cancer: American society of clinical oncology clinical practice guideline update. Journal of Clinical Oncology. 2013;**31**(19):2500-2510. DOI: 10.1200/JCO.2013.49.2678

[59] Anger JT, Gilbert BR, Goldstein M. Cryopreservation of sperm:indications, methods and results. The Journal of Urology. 2003;**170**(4 Pt 1):1079-1084. DOI: 10.1097/01.ju.0000084820.98430.b8

[60] Woods EJ, Benson JD, Agca Y, Critser JK. Fundamental cryobiology of reproductive cells and tissues. Cryobiology. 2004;**48**:146-156. DOI: 10.1016/j.cryobiol.2004.03.002

[61] Shehata F, Chian R-C. Cryopreservation of sperm and testicular tissue. In: Chian R-C, Quin P, editors. Fertility Cryopreservation. 1st ed. Cambridge: Cambridge University Press; 2010. pp. 39-50. Available from: http://www.thaisrm.com/docs/Ri%20Chian%20Book/Fertility%20preservation/whole%20book.pdf

[62] Isachenko V, Maettner R, Petrunkina A, Mallmann P, Rahimi G, Sterzik K, et al. Cryoprotectant-free vitrification of human spermatozoa in large (up to 0.5 mL) volume: A novel technology. Clinical Laboratory. 2011;**57**:643. Available from: https://pubmed.ncbi.nlm.nih.gov/22029178/

[63] Prentice JR, Anzar M. Cryopreservation on mammalian oocyte for conservation of animal genetics. Veterinary Medicine International. 2011;**2011**:146405. DOI: 10.4061/2011/146405

[64] Kuwayama M. Highly efficient vitrification for cryopreservation of human oocytes and embryos: The Cryotopmethod. Theriogenology. 2007;**67**(1):73-80. DOI: 10.1016/j.theriogenology.2006.09.014

[65] Arav A, Natan Y. The near future of vitrification of oocytes and embryos: Looking into past experience and planning into the future. Transfusion Medicine and Hemotherapy. 2019;**46**:182-187. DOI: 10.1159/000497749

[66] Dinnyes D, Dai Y, Jiang S, Yang X. High developmental rates of vitrified bovine cumulus oocytes following parthenogenetic activation, in vitro fertilization, and somatic cell nuclear transfer. Biology of Reproduction. 2003;**63**:513-518. DOI: 10.1095/biolreprod63.2.513

[67] Begin I, Bhatia B, Baldassare H, Dinnyes A, Keefer CL. Cryopreservation of goat oocytes and in vitro derived 2-4 cell embryos using the cryoloop (CLV) and solid-surface vitrification (SSV) methods. Theriogenology. 2003;**59**:1839-1850. DOI: 10.1016/s0093-691x(02)01257-8

[68] Parrott DMV. The fertility of mice with orthotopic ovarian grafts derived from frozen tissue. Journal of Reproduction and Fertility. 1960;**1**:230-241. DOI: 10.1.1.581.3573&rep=rep1&type=pdf

[69] Fuller B, Paynter S. Fundamentals of cryobiology in reproductive medicine. Reproductive Biomedicine Online. 2004;**9**:680-691. DOI: 10.1016/s1472-6483(10)61780-4

[70] Hashimoto S, Suzuki N, Yamanaka M, Hosoi Y, Ishizuka B, Morimoto Y. Effects of vitrification solutions and equilibration times on the morphology of cynomolgus ovarian tissues. Reproductive BioMedicine Online. 2010;**21**:501-509. DOI: 10.1016/j.rbmo.2010.04.029

[71] Nikiforov D, Russo V, Nardinocchi D, et al. Innovative multi-protectoral approach increases survival rate after vitrification of ovarian tissue and isolated follicles with improved results in comparison with conventional method. Journal of Ovarian Research. 2018;**11**:65. DOI: 10.1186/s13048-018-0437-5

[72] AbdelHafez FF, Desai N, Abou-Setta AM, Falcone T, Goldfarb J. Slow freezing, vitrification, and ultra-rapid freezing of human embryos: A systematic review and meta-analysis. Reproductive BioMedicine Online. 2010;**20**:209-222. DOI: 10.1016/j.rbmo.2009.11.013

[73] Bordes A, Lornage J, Demirci B, Franck M, Courbiere B, Guerin JF, et al. Normal gestations and live births after orthotopicautograft of vitrified–warmed hemi-ovaries into ewes. Human Reproduction. 2005:2745-2748. DOI: 10.1093/humrep/dei155

[74] Kagawa N, Silber S, Kuwayama M. Successful vitrification of bovine and human ovarian tissue. Reproductive Biomedicine Online. 2009;**18**(4):568-577. DOI: 10.1016/s1472-6483(10)60136-8

[75] Tada N, Sato M, Yamanoi J, Mizorogi T, Kasai K, Ogawa S. Cryopreservation of mouse spermatozoa in the presence of raffinose and glycerol. Journal of Reproduction and Fertility. 1990;**89**(2):511-516. DOI: 10.1530/jrf.0.0890511

[76] Nakagata N. Cryopreservation of mouse spermatozoa. Mammalian Genome. 2000;**11**(7):572-576. DOI: 10.1007/s003350010109

[77] Leibo SP, Sztein JM. Cryopreservation of mammalian embryos: Derivation of a method. Cryobiology. 2019;**86**:1-9. DOI: 10.1016/j.cryobiol.2019.01.007

[78] Terren C, Fransolet M, Ancion M, et al. Slow freezing versus vitrification of mouse ovaries: From ex vivo analyses to successful pregnancies after auto-transplantation. Scientific Reports. 2019;**9**:19668. DOI: 10.1038/s41598-019-56182-8

[79] Sztein JM, Takeo T, Nakagata N. History of cryobiology, with special emphasis in evolution of mouse sperm cryopreservation. Cryobiology. 2018;**82**:57-63. DOI: 10.1016/j.cryobiol.2018.04.008

[80] Nakagata N. Mouse spermatozoa cryopreservation. Journal of Mammalian Ova Research. 2000;**17**:1-8. DOI: 10/1274/jmor.17.1

[81] Takeo T, Nakagata N. Combination medium of cryoprotective agents containing L-glutamine and methyl-{beta}-cyclodextrin in a preincubation medium yields a high fertilization rate for cryopreserved C57BL/6J mouse sperm. Laboratory Animals. 2010;**44**(2):132-137. DOI: 10.1258/la.2009.009074

[82] Takeo T, Nakagata N. Reduced glutathione enhances fertility of frozen/thawed C57BL/6 mouse sperm after exposure to methyl-beta-cyclodextrin. Biology of Reproduction. 2011;**85**(5):1066-1072. DOI: 10.1095/biolreprod.111.092536

[83] Takeo T, Hoshii T, Kondo Y, Toyodome H, Arima H, Yamamura K, et al. Methyl-beta-cyclodextrin improves fertilizing ability of C57BL/6 mouse sperm after freezing and thawing by facilitating cholesterol efflux from the cells. Biology of Reproduction. 2008;**78**(3):546-551. DOI: 10.1095/biolreprod.107.065359

[84] Ostermeier GC, Wiles MV, Farley JS, Taft RA. Conserving, distributing, and managing genetically modified mouse lines by sperm cryopreservation. PLoS One. 2008;**30**:e2792. DOI: 10.1371/journal.pone.0002792

[85] Nishizono H, Shioda M, Takeo T, Irie T, Nakagata N. Decrease of

fertilizing ability of mouse spermatozoa after freezing and thawing is related to cellular injury. Biology of Reproduction. 2004;**71**:973-978. DOI: 10.1095/biolreprod.103.024422

[86] Nakagata N. Use of cryopreservation techniques of embryos and spermatozoa for production of transgenic (Tg) mice and for maintenance of Tg mouse lines. Laboratory Animal Science. 1996;**46**:236-238. Available from: https://pubmed.ncbi.nlm.nih.gov/8723247/

[87] Critser JK, Russell RJ. Genome resource banking of laboratory animal models. ILAR Journal. 2000;**41**:183-186. DOI: 10.1093/ilar.41.4.183

[88] Agca Y. Cryopreservation of murine oocyte and ovarian tissue. ILAR Journal. 2000;**41**:207-220. DOI: 10.1093/ilar.41.4.207

[89] Candy CJ, Wood MJ, Whittingham DG. Restoration of a normal reproductive lifespan after grafting of cryopreserved mouse ovaries. Human Reproduction. 2000;**15**:1300-1304. DOI: 10.1093/humrep/15.6.1300

[90] Jensen AK, Kristensen SG, Macklon KT, Jeppesen JV, Fedder J, Ernst E, et al. Outcomes of transplantations of cryopreserved ovarian tissue to 41 women in Denmark. Human Reproduction. 2015;**30**:2838-2845. DOI: 10.1093/humrep/dev230 ED 71

[91] Kato M, Ishikawa A, Kaneko R, Yagi T, Hochi S, Hirabayashi M. Production of transgenic rats by ooplasmic injection of spermatogenic cells exposed to exogenous DNA: A preliminary study. Molecular Reproduction and Development. 2004;**69**:153-158. DOI: 10.1002/mrd.20161

[92] Kohn DF, Clifford CB. Biology and Diseases of Rats. Laboratory Animal Medicine. 2nd ed. 2002:121-165. DOI: 10.1016/B978-012263951-7/50007-7

[93] Hirabayashi M, Kato M, Ishikawa A, Kaneko R, Yagi T, Hochi S. Factors affecting production of transgenic rats by ICSI-mediated DNA transfer: Effects of sonication and freeze-thawing of spermatozoa, rat strains for sperm and oocyte donors, and different constructs of exogenous DNA. Molecular Reproduction and Development. 2005;**70**:422-428. DOI: 10.1002/mrd.20223

[94] Nakatsukasa E, Kashiwazaki N, Takizawa A, Shino M, Kitada K, Serikawa T, et al. Cryopreservation of spermatozoa from closed colonies, and inbred, spontaneous mutant, and transgenic strains of rats. Comparative Medicine. 2003;**53**:639-641. Available from: https://pubmed.ncbi.nlm.nih.gov/14727812/

[95] Seita Y, Sugio S, Ito J, Kashiwazaki N. Generation of live rats produced by in vitro fertilization using cryopreserved spermatozoa. Biology of Reproduction. 2009;**80**(3):503-510. DOI: 10.1095/biolreprod.108.072918

[96] Roca J, Martinez S, Vazquez JM, Lucas X, Parrilla I, Martinez EA. Viability and fertility of rabbit spermatozoa diluted in Tris-buffer extenders and stored at 15°C. Animal Reproduction Science. 2000;**64**:103-112. DOI: 10.1016/s0378-4320(00)00185-8

[97] Lopez-Gatius F, Sances G, Sancho M, Yaniz J, Santolaria P, Gutierrez R, et al. Effect of solid storage at 15°C on the subsequent motility and fertility of rabbit semen. Theriogenology. 2005;**64**:252-260. DOI: 10.1016/j.theriogenology.2004.11.015

[98] Foote RH, Carney EW. The rabbit as a model for reproductive and developmental toxicity studies. Reproductive Toxicology. 2000;**14**:477-

493. DOI: 10.1016/s0890-6238(00)00101-5

[99] Okuda Y, Seita Y, Hisamatsu S, Sonoki S, Shino M, Masaoka T, et al. Fertility of spermatozoa cryopreserved with 2% acetamide or glycerol through artificial insemination in the Japanese white rabbit. Experimental Animals. 2007;**56**:29-34. DOI: 10.1538/expanim.56.29

[100] Holt WV. Fundamental aspects of sperm cryobiology: The importance of species and individual differences. Theriogenology. 2000;**53**:47-58. DOI: 10.1016/S0093-691X(99)00239-3

[101] Sawada Y, Chang MC. Motility and fertilizing capacity of rabbit spermatozoa after freezing in a medium containing dimethyl sulfoxide. Fertility and Sterility. 1964;**15**:222-229. DOI: 10.1016/s0015-0282(16)35171-8

[102] Moce E, Lavara R, Vicente JS. Effect of an asynchrony between ovulation and insemination on the results obtained after insemination with fresh or frozen sperm in rabbits. Animal Reproduction Science. 2003;**75**:107-118. DOI: 10.1016/s0378-4320(02)00228-2

[103] Fox RR, Burdick JF. Preservation of rabbit spermatozoa: Ethylene glycol vs. glycerol for frozen semen. Proceedings of the Society for Experimental Biology and Medicine. 1963;**113**:853-856. DOI: 10.3181/00379727-113-28511

[104] Eo Y, Kim SH, Bang SG, Oh MG, Park CH, Yoon JT. Effect of extenders with TCG and DMSO on the viability of rabbit sperm. Journal of Animal Reproduction and Biotechnology. 2019;**34**:100-105. DOI: 10.12750/JARB.34.2.100

[105] Domingo et al. Effect of glycerol, n,n-dimethylformamide and n-methyl-2-pyrrolidone on rabbit sperm stored at 4°C and 16°C. Animal Reproduction. 2019;**16**(4):887-894. DOI: 10.21451/1984-3143-AR2018-0100

[106] Carroll J, Gosden RG. Transplantation of frozen-thawed mouse primordial follicles. Human Reproduction. 1993;**8**:1163-1167. DOI: 10.1093/oxfordjournals.humrep.a138221

[107] Baird DT, Land RB, Scaramuzzi RJ, Wheeler AG. Functional assessment of the auto transplanted uterus and ovary in the ewe. Proceedings of the Royal Society B: Biological Sciences. 1976;**192**(1109):463-474. DOI: 10.1098/rspb.1976.0024

[108] Gray Mark E, James M, Paul S, Marland Jamie RK, Greenhalgh Stephen N, Rachael G, et al. Ovine pulmonary adenocarcinoma: A unique model to improve lung cancer research. Frontiers in Oncology. 2019;**9**:335. DOI: 10.3389/fonc.2019.00335

[109] Iris R, Baptista Pedro M, Anna L-C, Luca M, Marco P, Florien J, et al. Large animal models in regenerative medicine and tissue engineering: To do or not to do. Frontiers in Bioengineering and Biotechnology. 2020;**8**:972. DOI: 10.3389/fbioe.2020.00972

[110] Kopeika J, Thornhill A, Khalaf Y. The effect of cryopreservationon the genome of gametes and embryos: Principles of cryobiologyand critical appraisal of the evidence. Human Reproduction Update. 2015;**21**(2):209-227. DOI: 10.1093/humupd/dmu063

[111] Bailey JL, Bilodeau JF, Cormier N. Semen cryopreservation indomestic animals: A damaging and capacitating phenomenon. Journal of Andrology. 2000;**21**(1):1-7. Avaialble from: https://pubmed.ncbi.nlm.nih.gov/10670514/

[112] Rall WF, Fahy GM. Ice-free cryopreservation of mouse embryosat-196 degrees C by vitrification. Nature. 1985;**313**(6003):573-575. DOI: 10.1038/313573a0

[113] Massip A, Zwalmen PVD, Scheffen B, Ectors F. Pregnancies

following transfer of cattle embryos preserved by vitrification. Cryo Letters. 1986;**7**:270-273

[114] Széll A, Zhang J, Hudson R. Rapid cryopreservation of sheepembryos by direct transfer into liquid nitrogen vapour at -180degrees C. Reproduction, Fertility, and Development. 1990;**2**(6):613-618. DOI: 10.1071/RD9900613

[115] Fuku E, Kojima T, Shioya Y, Marcus GJ, Downey BR. In vitrofertilization and development of frozen-thawed bovine oocytes. Cryobiology. 1992;**29**(4):485-492. DOI: 10.1016/0011-2240(92)90051-3

[116] Vajta G, Holm P, Kuwayama M, Booth PJ, Jacobsen H, Greve T, et al. Open pulled straw (OPS) vitrification: A new way toreduce cryoinjuries of bovine ova and embryos. Molecular Reproduction and Development. 1998;**51**(1):53-58. DOI: 10.1002/(SICI)1098-2795(199809)51:1<53::AID-MRD6>3.0.CO;2-V

[117] Chasombat J, Nagai T, Parnpai R, Vongpralub T. Pretreatment of in vitro matured bovine oocytes with docetaxel before vitrification:effects on cytoskeleton integrity and developmental ability afterwarming. Cryobiology. 2015;**71**(2):216-223. DOI: 10.1016/j.cryobiol.2015.07.002

[118] Kristensen SG, Andersen CY. Cryopreservation of ovarian tissue: Opportunities beyond fertility preservation and a positive view into the future. Frontiers in Endocrinology. 2018;**9**:347. DOI: 10.3389/fendo.2018.00347

Chapter 3

Embryo Transfer

Ștefan Gregore Ciornei

Abstract

Assisted reproductive technologies (ART) have made tremendous advances, in last years. Artificial insemination is a method for achieving slow genetic progress in populations of animals. Many large and small ruminants are bred by AI, and more than a half million embryos are transferred every year around the world. Most of the ruminants sires used for artificial insemination were derived from embryo transfer. Improvements of reproductive biotechnologies of controlling the estrous cycle and ovulation have resulted in more effective programs for AI, superovulation of donor, and the management of ET. In the ruminants, ET procedure is a timely alternative that can allow good conception rates to be obtained constant in a year. There have been great advances of this biotechnique with on aimed to intensify the genetic progress between generations of farm. The gains is possible with the development of advanced reproductive biotechnique. The best current strategy in applying biotechnology to farmers is to use AI with sexed semen, so farmers will enjoy and benefit. The use of ET together with cryopreserved sexed embryos has a very specific potential for donor replacement and genetic improvement of the herd. In this chapter, procedures of the MOET protocol were described step by step.

Keywords: embryotransfer (ET), ruminant, cow, estrus, IVEP, MOET, ART

1. Introduction

The ruminants sector plays an important role in global socioeconomic directions. Therefore, it is necessary to research, to discover and to innovate and transfer knowledge to the farmer, for practices and alternatives that improve ruminants reproduction and production.

Increasing the efficiency of breeding and production of a farm herd is one of the great challenges for large ruminant producers. Recently and now, genetic selection programs have sought the characteristics needed to increase milk production, with gains by increasing the quantity and quality of milk. However, reproductive efficiency was neglected. In recent years, various publications have presented strategies to further increase milk and meat production and also to increase reproductive performance, which is a key factor for the efficient growth of ruminants [1].

In farmers, importance of a sustainable, economically viable production system can be obtained by maximizing reproductive efficiency of the ruminants herd.

This reproduction management can determine the profitability from the number of offspring produced, the genetic progress and the shorter interval between lactations periods.

The essential importance of precision reproductive care is therefore highlighted. This reproductive biotechnology (ET, AI) applied is capable of produce maximum production efficiency in animal farms with vulnerable populations, or in limited areas, in addition to improving animal welfare.

The first biotechnology of reproduction represented by artificial insemination (AI) is known as the simplest and lowest cost of reproductive biotechnology. This technique enhances the male's genetics, bringing slow genetic progress. However, dairy breeds in conditions of seasonality and climate change exhibit failures in estrus cycling and demonstration, which compromises AI results. In other words, with the discovery and description of FTAI protocols that synchronize follicular growth and induce ovulation, it is possible to achieve a high rate of inseminated animals without the need to observe the clinical and behavioral signs of estrus. [2], thus providing an increase in the conception rate and avoiding the occurrence of human errors in the detection of estrus, and calculating the optimal time for insemination. However, gametes an embryo can undergo degeneration in the extreme temperatures of summer [3]. The transfer of embryos produced in vivo (ET) became a strategy to avoid the deleterious effects of this period and provide a higher productive index than with AI [4].

With the beginning of the evolution of modern biotechnologies, the next step as major commercial progress in reproductive biotechnology was the transfer of embryos that appeared in the late 1970s. The ability to preserve, freeze and transport bovine embryos around the world has made ET an extremely useful technology for disease control, genetic rescue of valuable individuals and the development of new lines or breeds of animals.

ET is a multifactorial protocol that depends on several carefully and correctly performed sequential steps. Poor performance in any of the steps directly affects the success rate of the final result, the conception rate and the number of weaned products.

The use of embryo transfer as a breeding technique is growing throughout Europe, even in countries with less embryo transfer tradition. Historically the entire embryo transfer process was carried out at a specialist centres but now experienced reproduction vets are starting to carry out the artificial insemination (AI), flushing and searching as ambulatory procedures for transfer into a suitable recipient. The most time consuming and difficult part of the in vivo embryo transfer process is synchronizing recipients and transferring the embryo into the most suitable recipient. Receptors must be selected with with the best chance of maintaining the pregnancy [5, 6].

Embryo transfer provided a means by which the number of conception products could be multiplied rapidly, with the same origin. However, embryo transfer veterinarians have developed technology for commercial use and taken techniques from the laboratory to the farm. There have been countless practical difficulties for practitioners in uniting and setting up the International Embryo Transfer Society (IETS) in order to facilitate the discussions and steps deemed necessary for progress. Currently, the vast majority of countries in the European Union have one or more embryo transfer associations, where these actions are reported and come to support and develop ET biotechnology (eg AETE, SIET, AET-d, AETF, ARET and others).

Embryo transfer (ET) is now commonly used to produce AI sires from the top producing cows and proven bulls for the dairy industry [7]. As a perspective, the new genomic techniques presented are increasingly used for the selection of embryo donors, and genomic analysis has become essential for the selection of bull dams that will be used in embryo transfer [8]. Although the economy sometimes seems to exclude the use of embryo transfer techniques for anything other than gamete production, the commercial cattle industry benefits from the use of commercial males produced through well-designed MOET programs [9].

With the explosive development of this biotechnology, the techniques for obtaining embryos have been improved, the materials and consumables have become more efficient, the equipment more efficient, which makes the production cost of the embryo decrease and be higher quality. This desideratum is fully accepted by farmers and who apply this ET biotechnology more and more frequently [10].

Although there has been no appreciable increase the embryo production per poliovulated donor in last years, the importance of follicle wave dynamics [11] and methods for the synchronization of follicular wave emergence [12, 13], they simplified the protocols by which female poiliovulation could be achieved, leading to increased embryo production per application session. Currently, donor cows are hyper-stimulated more frequently than in the past (at an interval of 30–60 days) and thus more embryos can be produced per year, without changes in the current superstimulation protocol [14]. Other authors [15] have been interested in various factors that affect the viable production of embryos in animals and especially in dairy cows.

Potential embryo donors can be inseminated naturally or artificially (AI) and the embryos are normally collected non-surgically from 6 to 8 days after fertilization. After collection, the embryos must be identified and then evaluated morphologically. The evaluation procedure is done in an appropriate environment before the transfer. At this stage, they can be subjected to manipulations, such as splitting and sexing, and can be cooled or frozen for shorter periods or longer storage [16]. Discussion of donor superovulation, recipient synchronization, and embryo transfer should begin with a review of recent information on the physiology of female reproduction and the estrous cycle.

The reproductive genetic potential of every normal newborn calf is enormous. It is said that there are about 150,000 "eggs" or potential oocytes in a female and many billions of sperm produced by each male. We can say that through natural reproduction, only a small part of the reproductive potential of a valuable individual could be realized. The bull will be able to produce an average of 15 to 50 calves per year, and the cow will have an average calf per year. With the use of artificial insemination biotechnology, it is possible to exploit the large number of sperm produced by a genetically superior bull; however, the reproductive potential of the female with superior genetics was largely unused. It will produce on average 5–8-10 calves in its entire biological life through normal management programs. As artificial insemination has done for bulls, embryo transfer is a technique that can greatly increase the number of offspring that a genetically important cow can prove and produce. The main reason for the development of embryo transfer in cattle was to further the increase in genetic progress of the female.

2. Advantages of embryo transfer (ET)

- Increase the small population of valuable animals [17, 18].

- Helps in the genetic improvement of animal, decreases the generation interval, increases selection intensity [19].

- It helps to get more many calves from a genetically superior single female as against a few calves which can be produced naturally in its lifetime

- An possibilites tool to produce breeding bulls from a limited number of superior females for use in AI [20].

- May increase the numbers of the existing purebred herd.
- Possible to obtain offspring from the genetically valuable female that has become infertile due to disease, injury, or age [21].
- With the improvement in cryopreservation and the advancement of technologies, it is economical and easier to transport (import/export) embryos instead of living animals.
- Endangered animals can be saved from extinction by embryo production and cryopreservation [22].
- Avoids transmission of diseases from infected donors to their offspring-Genetic defects [20].
- Decreases the risk of transmitting infectious agents. Embryo collected from cows with bovine leukemia virus, blue tongue virus, FMD virus if washed properly before transfer (Trypsin treatment) to unaffected recipients does not transmit disease.
- ET allows differentiation between normal and abnormal fertilization [23].
- An important tool for disease control, genetic salvage of valuable animals, biosecurity program, development of new lines/breeds of animal.
- Helps in the proliferation of femele genetic material from dam and sire [24].
- Associated reproductive technologies (ART) such as embryo splitting, sexing of embryos, cloning, transgenesis further broadens the horizons of ET [25].
- The males produced by ET out of superior donors by using elite bull semen can be used as future elite sires on the AI network and females produced would serve as the future bull mother. Therefore the existing demand of elite sires especially in cattle and buffalo breeding programs can be largely met [24].
- Can be used to evaluate the contribution of the aging oocyte to decreased reproduction in geriatric animals [23].
- Infertility treatment, ET serves as an important tool in the treatment of infertility in cows.

3. Sexual cycle physiology

The intrinsic control of the bovine estrous cycle is coordinated by the interdependent secretion of hormones from structures such as: hypothalamus, anterior pituitary, ovaries and uterus [26]. These include gonadotropin-releasing hormone (GnRH) from the hypothalamus, folliculostimulating hormone (FSH) and luteinizing hormone (LH) from the anterior pituitary gland, estrogen, progesterone and inhibin from the ovary and prostaglandin F2a (PGF) from the uterus.

During gestation, the multiplication phase of the oogonia results in the constitution at the birth of a stock of primordial follicles, the number of which in the cow is between 200 and 250,000. These primordial follicles have a diameter of between

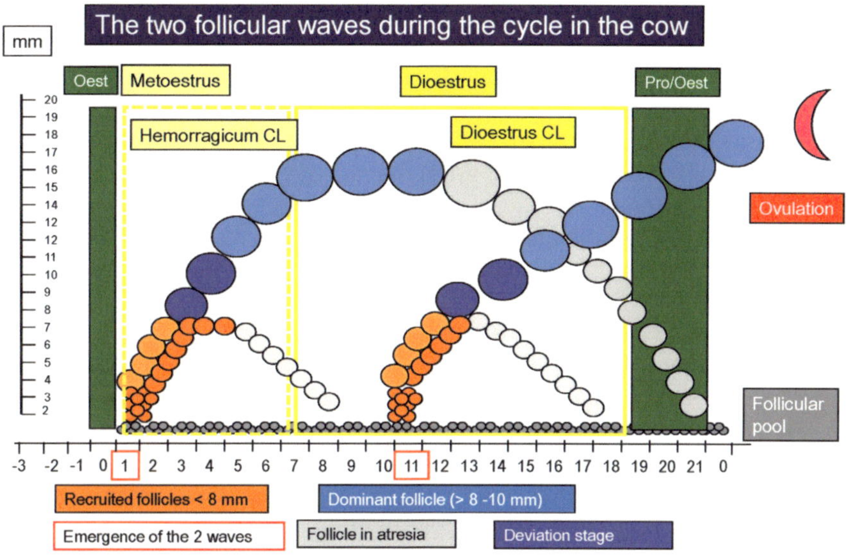

Figure 1.
Physiology of the sexual cycle in cows, with two follicular waves.

30 and 40 microns. At the secondary and especially tertiary stage, a cavity appears. It then becomes possible to identify by ultrasound these follicles with a diameter of between 2 and 4 mm.

In cows, as in many other mammals, follicular growth develops in the form of waves. In a 21-day cycle, there are two waves. The presence of a third wave is not uncommon. It has the effect of lengthening the average length of a cycle by a few days (24 vs. 21 days). When the cycle has two waves (**Figure 1**), the emergence of one wave occurs on days 0–1 and 10–11 of the cycle. Day 0 corresponds to that of estrus. By emergence we mean the moment or by ultrasound, it is possible to distinguish in the mass of follicles recruited the one which will become dominant. During the 2.5 days after the emergence of a wave, the selected future dominant and dominated follicles continue to grow. The dominant follicle reaches at this time the average diameter of 8 or even 9 mm. This moment is called "follicular deviation" and characterizes the moment when the dominant follicle will be able to clearly distinguish itself from other growing follicles. Its diameter is therefore 2 mm greater than that of the other selected follicles. The follicle continues to grow until it reaches a diameter of 10 mm. Clinically, this dominance can be identified by ultrasound, or by Doppler ultrasound to identify changes in the vascularization of the follicle or by assaying hormones such as estradiol or follicular fluid inhibin. This dominance is therefore both morphological, ie exerted by the largest follicle and physiological, because it brings about an arrest in the development of the dominated follicles which go through a static phase before settling down. This physiological dominance also implies the appearance at the granular level of LH hormone receptors which will take over from FSH to ensure further growth of the dominant follicle. The period of physiological dominance is shorter than that of morphological dominance. Clinically, the identification of more than 10 follicles with a diameter of between 3 and 8 mm makes it possible to exclude the presence of a physiologically dominant follicle. Growth of the dominant follicle will continue until it reaches a maximum diameter of 16 mm.

This is followed by a so-called static plateau phase lasting more or less 6 days at the end of which the dominant follicle will begin to regress. Due to the gradual decrease in estradiol synthesis by the dominant follicle, there is a new release of

FSH and the appearance on day 10 of new follicular emergence. This cannot be observed as long as the dominant follicle from the previous wave is in the growth or plateau phase. This new wave develops like the previous one through the dominance of a new follicle which will suppress the growth of the subordinate follicles which will become saturated. The follicle continues to grow. In proestrus he is no longer under the progesterone influence, so given the massive release of LH he can stop growing and then ovulate.

During the cycle, the follicular population is therefore distributed into several classes except that of the follicles in the reserve. A first class consists of recruited follicles. Their diameter is 2 to 4 mm. A second class is made up of growing follicles. These follicles can potentially become the ovulatory follicle. Their diameter is between 6 and 10 mm. The third class refers to the dominant follicle. Its diameter is between 10 and 16 mm. Finally, can we also identify the preovulatory follicle with a diameter greater than 15 mm. It will persist on the ovary for 5 to 6 days before regressing or ovulating (**Figures 1** and **2**) [5, 6, 10, 27].

In a cycle with three waves (**Figure 2**), the emergence of cows occurs on days 1, 10 and 17 respectively, with day 0 being estrus and therefore day 1 ovulation. The general wave pattern is comparable to that described for a cycle with two waves. Waves 1 and 2 are anovulatory. Only the third is normally ovulatory. It will be seen that the luteal phase like the cycle is of longer duration than for a cycle with two waves. Likewise, the interval between the onset of the ovulatory wave and ovulation is shorter (7 vs. 11 days).

The main mechanism of synchronization of the estrous cycle is ovulation, when the first follicular wave occurs [11]. A new hormone-secreting endocrine gland is formed instead of the ovulatory follicle and is called the corpus luteum (CL) it is formed in the following days (3–5) and in the absence of pregnancy, it wraps around day 16 or 17 of the cycle [26]. The most common hypothesis for CL regression is that the non-pregnant uterus secretes a luteolytic factor into the uterine venous blood. This substance is transferred through a local blood (veno-arterial) pathway to the ovarian artery through which it reaches the ovary and causes luteolysis [27].

After CL regression, a rapid decrease in serum progesterone concentrations to values lower than 1 ng/ml results, at the same time, the frequency of LH pulse increases and follicular development is further stimulated. The growth and maturation of the follicle that becomes preovulatory results in increased estradiol secretion, which causes local estrogenic changes in the oviduct and uterus, behavioral estrus and a preovulatory release of LH (around the time of estrus manifestation). The preovulatory LH peak results in the resumption of the oocyte meiosis process, and ovulation 24 to 32 hours later and the luteinization of the ovulated follicle to form a secretory corpus luteum hemoragicum. The growth and development of the hemorrhagic corpus in a fully functional CL results in progestative changes in the oviduct and uterus that are favorable for embryonic development and pregnancy. If pregnancy does not occur, the cycle resumes again with the disappearance of CL [6, 11, 23, 27].

3.1 Estrus synchronization, superovulation

Estrus synchronization and superovulation are critical components of an embryo transfer program. These techniques involve the manipulation of the basic endocrine patterns, presented and described in this document [28]. The key to successful estrus synchronization is synchronous growth and ovulation of a viable dominant follicle and closely synchronized, rapid declines in circulating progesterone to values <1 ng/ml [29]. If properly implemented, within the physiological constraints of their mechanism of action, current techniques for synchronization of estrus and ovulation are highly successful [30]. However, the variation in the

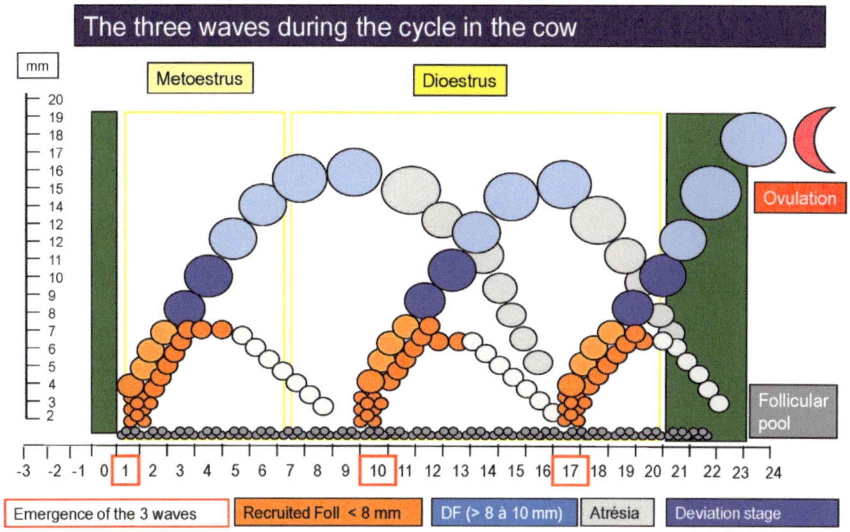

Figure 2.
Physiology of the sexual cycle in cows, with three follicular waves.

dynamics of ovarian follicular waves makes it difficult to control the exact time of estrus and ovulation.

The goal of superstimulation treatments in cows is to obtain the maximum number of fertilized and transferable embryos with the highest possible probability of inducing and sustaining a pregnancy.

The variability of the ovarian response was related to differences in superstimulatory treatments, such as gonadotropin preparation, gonadotropin type, duration of treatment, timing of previous estrus treatment, total gonadotropin dose, and use of additional hormones in superstimulation. Protocol [4]. Additional, equally important, sources of variability are factors inherent in the animal and its environment. These factors may include nutritional status, reproductive history, age, season, breed, ovarian status at the time of treatment and perhaps most importantly, inherent numbers of antral follicles [5]. While considerable recent progress has been made in the study of bovine reproductive physiology, factors inherent to the donor animal that affect superovulatory response are only partially understood [13, 25, 30].

4. Embryo transfer procedures

In farm animals, fertilized ova is removed from the uterus of their dam (the donor) and transferred to the uterus of other females (recipients) for development to term. Almost all commercial embryo transfers now use nonsurgical methods to recover the embryos rather than surgical methods (only for small ruminants). The procedure requires multiple steps (**Figure 3**), a large amount of time, and a variable cost.

The stages of a direct/in vivo ET protocol are highlighted in the following mandatory steps [31]:

1. Donor cows, selection of embryo donors.

2. Poliovulatory treatment of donors,

3. Artificial insemination/mounting,

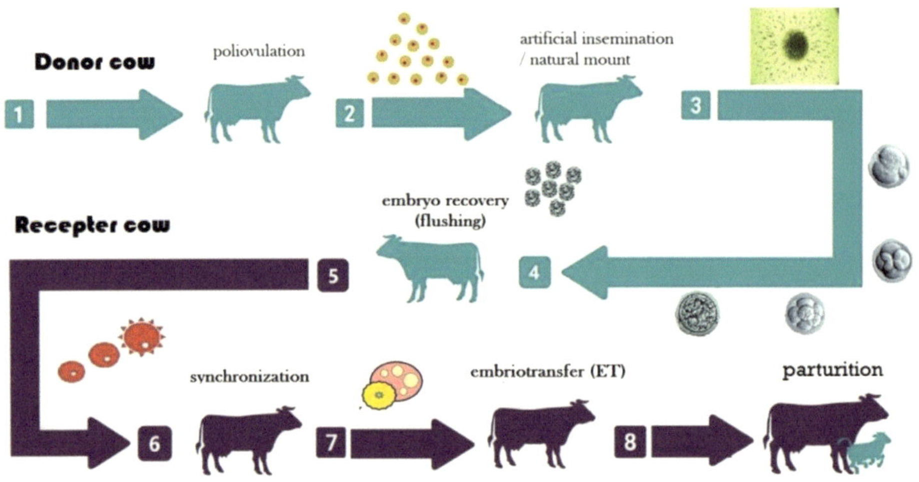

Figure 3.
Stages of in vivo embryo transfer in large ruminants.

4. Collection of embryos and classification,

5. Selection and preparation of receptors,

6. Synchronization of estrus and ovulation with the donor,

7. Direct transfer/preservation,

8. Gestation and parturition.

4.1 Donor selection

The selection of the embryo donor candidate is based on two major criteria: (1) the genetic merit, generally evaluated by the owner and based on performance, and (2) the reproduction criteria interpreted and evaluated by the veterinarian. The donor must be in good physical condition, an average but growing BCS. It should be free of underlying conditions, be at least 50 to 60 days after calving and have a regular cycle. In general, cows with a history of reproductive problems, even minor ones, do not make good embryo donor animals.

Donors are further evaluated by careful examination of the cervix, uterus, and ovaries per rectum to determine if they lack adhesions to neighboring organ structures, and the presence of other palpable lesions. It is recommended to test the permeability of the cervical canal with a cervical dilator, especially if the donor is before the first calving - heifer. This prevents the occasional of being unable to negotiate the cervix after a series of costly hormonal injections.

Single or multiple embryos can be collected from ovulating or naturally superovulated cows. For optimal efficiency, 2 to 4 donors should be treated and synchronized with their recipients for each attempt/session; this allows the sharing of the recommended potential of 8–10 recipients per donor.

4.2 Superovulation

Superovulation is and remains one of the least anticipated steps in the process of embryo production. The objective of superstimulation treatments in the cow is to obtain the maximum number of fertilized and transferable embryos with a high probability of producing pregnancies [32].

In the bovine tremendous variation in response occurs with age, breed, lactational status, nutritional status, season, and stage of the cycle at which treatment is initiated. Follicle stimulating hormone (FSH), which has a short half-life (Pluset, Folltropin-V, and others), necessitates twice-daily injections over a period of 4 to 5 days. Synthetic hormones with a long half-life (like PMSG), are administered in a single dose, but have other drawbacks. Treatment is start in the mid-luteal phase (day 8 to 12) of the donor's estrus cycle and white use of prostaglandins (PGF) to synchronize the estrus of the donors and the recipients. Alternatively, treatment may be induced on day 16 or 17 (day 0 = estrus) of the donor's natural estrous cycle, or with progesterone administration (which mimics a progesterone phase). Ultrasonography and palpation of the ovaries per rectum have been shown to have similar accuracy for determination of the number of follicles (in estrus fase) or CL (at the time of embryo recovery). However, the number of anovulatory follicles can be counted more accurately [1, 4, 33, 34].

4.3 Artificial insemination/mounting

Donors should be artificially inseminated twice with a 10–12 hour interval, beginning 6 hours after the occurrence of oestrus, to ensure the time interval in which ovulation occurs. Depending on the quality of the frozen/sexed semen, a dose with a higher sperm concentration, even a double dose, can be used for each insemination. Ultrasonography is helpful in assessing the potential superovulatory response on the day prior to ovulation or at the time of AI.

4.4 Collection of embryos and classification

The donor animal is kept in a standing position in a trevis. The first step in the non-surgical recovery of embryos is to determine the numbers of corpus luteum in the ovary [21]. This step is important to rule out that the superovulation response in the animal; if less or no CL is found-indicates the poor response of superovulation-flushing not to be done in such animals. The donor was given an epidural anesthesia, then a wash and disinfection of the ano-vulvar region (**Figure 4**).

A two-way round tip balloon catheter (Fr. size 16 to 24) with a tul inflatable balloon is used. Once the instruments has been made ready (two/tree-way catheter), the vulvar lips are parted and the catheter with stylet is inserted into the vagina and advanced towards the lumen of the cervix. It is further advanced to the horn of the uterus until the balloon is situated at the base of the uterine horn. By blowing air, a dam is created with the tip of the uterine horn, there are located the embryonic formations between days 5–8 after ovulation. The amount of air used depends upon the size of the uterus. Basically, there are two methods of embryo collection [35]: the continuous or interrupted flow, closed-circuit system and the interrupted-syringe technique. The most commonly used medium for embryo recovery is Dulbecco's phosphate buffered saline (PBS), but there are many others ready to use (Euroflush, Vigro). Uterine horn is flushed with 30–60 ml of media and repeated until 300–800 ml of media is used up. The same process is repeated for the other horn as well [36].

Embryos are found with a 10 X magnification stereoscope after filtering the collection/washing medium through a pore filter with a diameter between 50 and 70 μm. The identified embryos are usually transferred as soon as possible, sometimes if desired it is possible to keep the embryos in that environment for a few hours at room temperature. It is also possible to cool the bovine embryos in storage

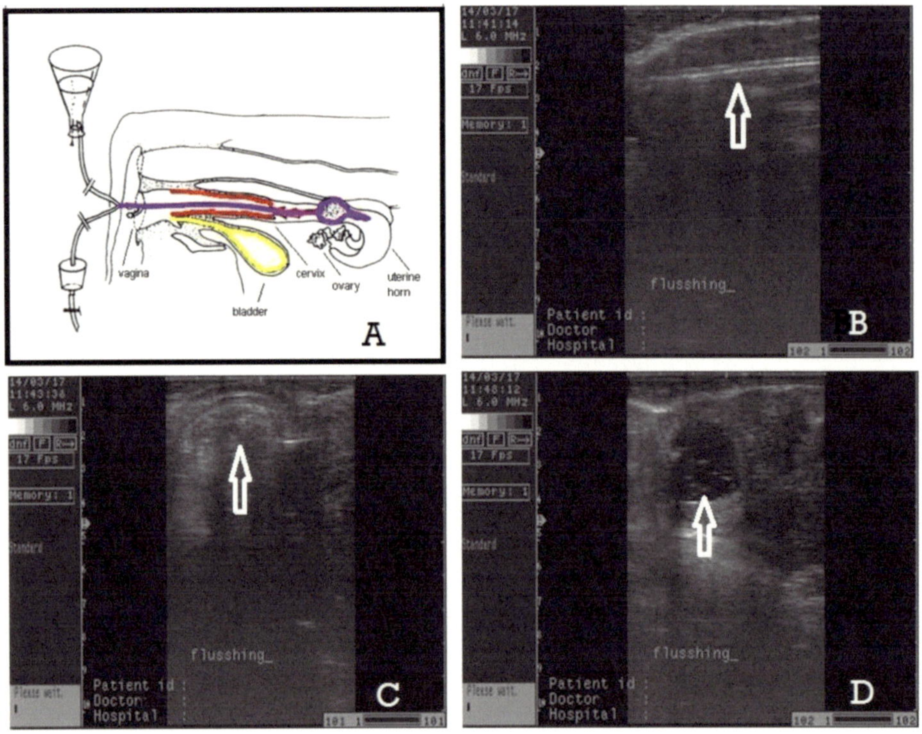

Figure 4.
Recovery of bovine embryos by flushing method. A. Scheme of catheter placement and reservoir-uterus-filter fluid flow, ultrasound images with catheter guidance through the cervix and visualization of uterine lavage, see white arrow. B. Inserting the stylus through the cervical lumen, C. inflating the balloon from the catheter and obtaining the dam at the top of the uterine horn, D. flow of flushing fluid and recovery of embryos.

medium and store them in the refrigerator for 2 or 3 days. Most often, embryos can be frozen for use at a later date.

A good response and an appropriate recovery rate results in getting a 4–5 embryos are recovered with each flush. This can lead to 50 freezable embryos per donor per year resulting in the birth of 30 calves after the transfer of the embryo to a recipient [24].

After the fecundation, the single-celled embryo now called the zygote undergoes rapid mitotic divisions (cell number increases, cytoplasm remaining same) called cleavage [37]. Bovine embryo descends into the uterus around day 4.5 days (estrus day 0) [38]. According to the standards, embryos are recovered from six to eight days after the onset of estrus (day 0). Embryos can be recovered even earlier from four days when the embryos arrive from the salpinx in the uterus, but before day 6 the recovery rates are lower than on days 6–8 (**Table 1**).

Species	Days from estrus
Cattle	7
Buffalo	6
Sheep	3–6
Goat	3–4

Table 1.
Day of collection of embryo.

However, embryos can also be recovered on days 9–14, although they leave the pellucid area on days 9–10, making them more difficult to identify and isolate from cellular detritus and more susceptible to infection [39].

Identification and evaluation of embryos is one of the most important and delicate stage, the practitioner needs experience to get used to the procedure. Embryo quality and poor handling techniques can directly affect pregnancy rates. A step-by-step procedure for looking for embryos is presented in the content of this section.

Evaluation of the embryo in the uterine effluent is based on identification of several morphologic features of the embryo using light microscopy. These methods are subjective and depend on experience. The embryo is spherical and is composed of blastomeres surrounded by a gelatin-like shell and zona pellucid (**Figure 5**).

Embryos recovered 5 to 8 days after estrus are classified morphologically into the following groups, based on their stage of development. Proper evaluation requires rolling of the embryos along the bottom of the dish.

Morula. The blastomeres are round and not tightly joined together. Individual blastomeres are difficult to identify with each other. The blastomere cell mass of the embryo occupies most of the perivitellin space.

Compact morula (tight morula). The shape and appearance of a tight mill is similar to a golf ball, in that the outer edge is slightly wavy (curly) given due to compaction. Individual blastomeres grow and become indistinguishable. The cells on the surface of the mass have a polygonal shape. The embryonic cell mass occupies 60–70% of the perivitellin space.

Early blastocyst. A tiny transparent (clear) space that contains fluid is visible. This area is the beginning of the blastocele (cavity). The embryo occupies 70–80% of the perivitelline space [6, 14, 37].

Blastocyst. The blastocele cavity becomes prominent and represents more than 70% of the embryo's volume. Inside, two groups of cells are separated and differentiated. They can be clearly recognized as a trophoblastic layer below the pellucid area and the darker cell mass occupying part of the embryo. The perivitellin space is still visible, but it is very small.

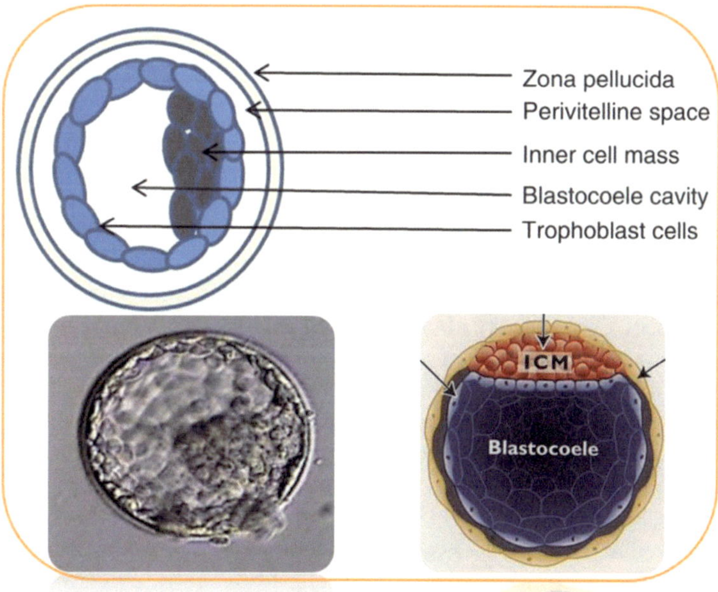

Figure 5.
Schematic diagram of a transferable embryo (expanded blastocyst phase).

Expanding or expanded blastocyst. There is no more perivitelline space between the trophoblastic cell layer and the interior of the area. The area of the pellucida stretches becomes thinner as the blastocyst expands. A small, well-compacted internal cell mass is observed positioned in one part of the embryo. The color of the embryo becomes pale to clear and is due to the large amount of fluid present inside.

Collapsed blastocyst. A perivitelline space can be identified together with a very thin pellucid area. The blastocyst may be partially collapsed, with a smaller blastocyst cavity, or completely collapsed and have the appearance of a compact morula.

Hatched blastocyst. After a continuous growth, the blastocyst expands to rupture and the embryo escapes from the pellucid area. From this moment the embryos pass into the gastrulation phase. The hatched blastocysts can be spherical with a well-defined blastocyst or they can be collapsed, similar to cellular detritus. Identifying embryos at this stage is especially difficult for the inexperienced operator.

When blastocysts/gastrules without areas or hatching are collected, there is a higher risk of damage due to handling. In addition, hatched blastocysts are sticky and can adhere to micropipette handling tubes. Therefore, the use of embryonic filters is not recommended when there is a suspicion that hatched embryos will be recovered (> day 7.5).

Embryos are then classified according to quality based on morphologic appearance. Excellent/good, fair, and poor quality embryos are considered transferable into recipients. Excellent or good quality embryos (Code 1) are freezable (**Figure** 5).

4.4.1 Codes for embryo quality

Code 1: ***Excellent or good***. The mass of the embryo is symmetrical and spherical with individual blastomeres (cells) they are uniform in size, color and density. The embryo was in accordance with the expected stage of development (collection day). The irregularities are usually minor and more than 85% of the cellular material should be a compact and intact embryonic mass. This is based on the observation of the percentage of embryonic cells represented by the extruded material in the perivitelline space. The pellucid area should be smooth and smooth and could adhere to a micro-plate or a straw.

Code 2: ***Fair***. Some irregularities can be observed in the general shape of the embryo mass or in the size, color and density of individual cells. At least 50% of the cellular material must be an intact, viable mass of embryos.

Code 3: ***Poor***. Some major irregularities in the shape of the embryo mass, or the size, color and density of blastomeres, are identified. At least 25% of the cellular material should be like an intact, viable mass of embryos.

Code 4: ***Dead or degenerating***. Degenerate embryos, oocytes, or I-cell embryos are nonviable.

Embryos of fair quality can be transferred fresh, if the receptors are available and synchronized. The category of good and excellent quality embryos have a high probability of surviving cryopreservation. The EITS considers that the export of embryos of poor and fair quality is inadequate [40]. The assessment of bovine embryos has recently been revised and is constantly improving [41], but the IETS manual has the most comprehensive library of embryonic images useful to practitioners.

Loading the Straw. Immediately before direct transfer, the embryos are individually aspirated into sterile 0.25 ml French straw. The embryo is usually loaded from the support vessel into the straw using a 1 ml syringe attached to the end of the straw stopper. First, a 3 cm medium column is aspirated into the straw, then a 0.5 cm air column is aspirated, then a 3 cm medium column containing the embryo, followed by another air bubble.

Selection and preparation of receptor. The recipient should be non-pregnant, cyclic (minimum of two normal cycles), should have CL on at least one of the ovaries. The embryo stage should match the uterine age of the recipient for a successful pregnancy to occur. While perfect synchrony is desirable, recipients that are 1 day out of phase can be considered acceptable; this means that a 7-day embryo can be transferred into a recipient who showed heat 6–8 days earlier. The lower quality embryo is more sensitive to asynchrony. The recipient should not have any pathological condition which can hinder its pregnancy. Pregnancy rates following embryo transfer are best when the recipient is in estrus from 36 hours before to 12 hours after the donor [42].

Synchronous recipients can be produced in three ways:

1. Selection from a large pool of cycling females. This strategy limits the number of embryos and time when embryos can be collected. Approximately five percent of the herd will be in heat on any given day.

2. Estrous cycles of any number of recipients can be synchronized with PGF2alpha or its analogues, or with CIDR devices, to exhibit heat the same day as or just ahead of the donor.

3. Timed ET, analogous to timed AI (Ov-Sync), can also be used. The importance of close synchrony between the age and the stage of development of the embryo, and the endocrine status of the endometrium of the recipient must be emphasized. Pregnancy rates following embryo transfer are best when the recipient is in estrus from 36 hours before to 12 hours after the donor [43].

Embryo transfer to the recipient can be done surgical or non-surgically. However non-surgical is more ethical to use. The recipient is secured in a Travis and the vulvar area is cleaned. As the animal is in the luteal phase. Epidural anesthesia is induced to prevent straining and defecation. The insertion of the tip of the instrument into the desired uterine horn should be done quickly, and smoothly. Trauma to the delicate endometrium causes bleeding, and blood (complement in the serum) is embryocidal. Ruminants embryos are transferred to the uterine horn and the same procedure as A.I. is followed except that in ET embryos are deposited deep in the horn ipsilateral to CL [44].

Pregnancy rates for IVP embryos were lower in commercial embryo transfer programs than for in vivo embryos [45].

Pregnancy rates are 10% lower in frozen embryos than the fresh ones [37]. Using heifers as recipients, there have been reports that in some 10% of such animals (heifers) it is difficult, if not impossible, to carry out ET via the cervix.

Any kind of stress to the recipient should be avoided. Any other routine treatments scheduled (eg antiparasitic) must take place at least 3 weeks before the transfer; also changes in the feeding regime should be prohibited for 3–4 weeks before and after embryo transfer. Beneficiaries must be accommodated where they can be easily and quietly handled on the day of transfer [23]. Any stressors should be removed.

5. Embryonic mortality

It is said that about 25–40% of embryonic losses are produced in the first few days after transfer to the cow [46, 47]. It has been observed that most of these females return to heat at an interval after 20–22 days, presenting a complete and normal sexual cycle [48]; Therefore, it is believed that embryonic mortality (EM) could occur between days 7 and 17, the period from embryo transfer (ET) until it settles at maternal recognition of pregnancy [49]. In a lower proportion, but just as important, is the pregnancy losses that occur between days 28 and 98, after the transfer and the percentages between 7% and 33% have been reported [50].

The critical nature of the period and the phenomenon of recognition and survival of the embryo at the maternal uterine endometrium during implantation requires a very careful synchronization between the transferred embryo and the recipient. Thus, the importance of both the biochemistry of the uterine environment and the signals of the embryo that generates the recognition and implantation is highlighted [51, 52]. These embryonic signals must be released at the time and concentration necessary to maintain CL morphology and maintenance of function, thus generating a continuous production of P4. Progesterone levels play an essential role in maintaining the embryotrophic environment and supporting the normal development of the concept (the embryo and all adjacent cell structures) [48].

In connection with the influence of P4 (progesterone) on certain events related to pregnancy maintenance from the early stages and the ability of PGF2α to trigger luteolysis, a number of hormonal strategies for maintaining pregnancy have been researched, developed and supported [53, 54]. These strategies tend to be based on the increased efficiency and secretion capacity of P4 by CL: secretion must occur in a timely manner, thus ensuring a suitable uterine environment for the development of the embryo transferred to the recipient bovine female. All these strategies aim to increase the load rate in ET programs [48].

In order to prevent the mortality of the transferred embryos, and the loss of the pregnancies during embryo transfer sessions, two main objectives are considered: - Maintaining the corpus luteum function, even inducing a new one; and Inhibition of the appearance/secretion of luteolytic factor. All procedures apply to female embryo recipients.

In the first case, it is recommended to administer a treatment with Gn-RH, more precisely HCG to develop and support the luteal tissue, or even to form another CL (by causing ovulation of the follicle, if any). In the second case, the administration of non-steroidal anti-inflammatory drugs is considered, which is said to block the synthesis of PGF.

6. Embryo production biosecurity and contamination risks

The procedures for embryo production, in MOET programs, include several steps where contamination with pathogen agents may occur. For instance, the first source of potential contamination comes from the donor itself. Before ovulation, an oocyte could be contaminated by its contact with a given pathogen shed in granulosa cells or follicular fluid during infection (viremia or bacteremia). For example, in bovines, viruses were detected in follicular fluid a few days after experimental exposure to bovine viral diarrhea virus [55]. Hence, the recovery of cumulus–oocyte complexes at this moment might lead to production of contaminated embryos [6, 37].

Disease Risk Management. Success in embryo production by either MOET or IVEP relies on the capacity to correctly perform all technical steps, eliminating or reducing factors recognized to have negative effects. It is essential to select donors and recipients with good general health and adequate nutrition. In addition to those issues, considering that the first source of potential contamination comes from the donor itself, an important measure is to select these females, taking into account their sanitary status. When incorporating animals into the flock, their health status should be checked before and quarantine should be respected. Vaccination and deworming must be employed as prescribed, depending on the location and system of production, but always before their use as donors. Testing should be conducted for some infectious diseases, and those positive should be culled. All technicians in direct contact with the animals must be careful and well trained to ensure familiarity with and safety in the procedures. The technique must be aseptic and all labware sterile. The equipment should be cleaned and all devices that are in contact with the animals should be sterilized before reuse. Clothing should be completely cleaned before reuse [56].

In general, in IVEP, the risk of potential hazards associated with oocyte collection from slaughterhouses are higher than those collected by laparoscopic ovum pickup. Consequently, when using these ovaries, it is important to determine their origin, particularly whether ovaries came from a herd depopulated for any health cause [57]. Care must be taken in the transportation of this material to the laboratory to avoid any external contamination. For media preparation and gamete or embryo manipulation or culture, all biological products should be avoided. These reagents could be replaced by those derived from plant origin or amino acids. When cell culture is preferred for IVEP, the use of controlled cell lines, confirmed to be pathogen-free, is recommended. From a sanitary point of view, safer strategies include the use of chemically defined media that do not contain serum or somatic cells [57].

For MOET, pathogens could be present in the female genital tract and can adhere to either oocytes before fertilization or embryos before collection. Intact zona pellucida is a natural barrier to penetration of pathogen into the oocyte or prehatching embryos. However, some pathogens may adhere to the zona pellucida of oocytes and embryos; thus, the zona pellucida represents a vector for disease transmission to recipients and to embryos after hatching (once transferred). For IVEP, the magnitude of this risk may vary according to the source of ovaries or oocytes that are being used: either from laparoscopic ovum pickup when the donor health status is well known or from the slaughterhouse [57].

Follicular aspiration by laparoscopy, instead of transvaginally, practically eliminates the chance of contamination by microorganisms being carried into the follicle from the vagina via the collection needle, as has been reported in humans [58].

On the other hand, ovaries collected from slaughterhouses provide an inexpensive and abundant source of oocytes, which is usually helpful for research projects and cloning. However, considering that these ovaries are generally transported in containers together, the presence of just a few ovaries from infected animals could represent a potential source of contamination. Other general sources of possible contamination involve the presence of environmental pathogens associated with the technician, slaughterhouse, equipment for laparoscopic ovum pickup or embryo collection, or even in the laboratory, such as glassware, culture dishes, media, and incubators. Regarding media, it is known that any biological product such as fetal calf serum and bovine serum albumin used in the recovery, culture, and cryopreservation of oocytes, sperm, and embryos may constitute a risk of contamination [6, 37, 55–58].

The semen used in a ET protocol (MOET or IVEP program) must be collected from males managed under appropriate sanitary protocols that ensure their good health status. Although AI represents a useful tool for disease control when best practices are applied, a further source of risk in an embryo production program is the semen. Numerous viral, bacterial, and parasite agents may be present in semen, which may adhere to the surface of spermatozoa or they could be present in the seminal fluid or in the semen extender.

In general, the studies are in agreement when the sanitary procedures recommended by IETS are correctly implemented. The risk of disease transmission from donor to recipient and to offspring for most pathogens is negligible or, at least, is much lower than that associated with live animals. These facts confirm that embryo transfer represents a safe strategy for global genetics trade and a valuable tool for the control and eradication of several diseases in small ruminants [59, 60].

Various publications [57] describe the possibility of transmitting diseases and the management of prevention procedures. The procedures for managing these risks have been described in the ***OIE Terrestrial Code*** [61], which explicitly refers to the specifications published in the IETS manual. These procedures are included in national regulations for the transfer of embryos between herds.

Adherence to these procedures ensures that embryo transfers contribute to improving the animal health of a population by controlling the movement of genetic material between herds. The basic concept behind these regulations is the official validation of embryo transfer teams. This approval is a very important method of veterinary regulations, as they are usually based more on animals coordinated in protocols and their products. However, in this case, the safety of embryo transfer procedures is based on the correct ethics and technique of the head of the embryo transfer team [62].

The criteria used by national veterinary services for the approval of embryo transfer teams are based on the *Terrestrial Code*. For example, in Chapter 4.7. it is stated that: "the embryo collection team is a group of competent technicians-operators, including at least one veterinarian, who carry out the production, collection, processing and storage of embryos".

It is recommended that the following conditions be met:

a. the team should be approved by the competent national authority;

b. the team should be supervised and ordered by a team veterinarian;

c. the team veterinarian should be responsible for all operations of his team, including:

- checking the health of the embryo donor
- implementation of appropriate disease control measures when handling or operating donors
- disinfection and hygiene procedures;

d. team staff should be regularly trained appropriately in disease control techniques and principles. High standards of hygiene must be practiced to prevent the spread of infection;

e. the embryo collection team should have adequate equipment for:

- induction and collection of embryos
- processing and treatment of embryos in a permanent laboratory or in a mobile laboratory
- conservation and storage of embryos;

These facilities do not necessarily have to be in the same location;

f. the embryo collection team must draw up a correct record of its sessions, which should be kept for verification by the Veterinary Authority for a period of at least two years after the export/movement of the embryos;

g. the embryo collection team should be periodically inspected and checked at least once a year by an official veterinarian, to ensure compliance with the procedures and sessions for the proper collection, processing and storage of embryos.

7. Conclusion

Embryotransfer *in vivo*, (IVD by MOET) is a procedure that can significantly increase the amount of offspring a genetically significant ruminants can carry. This it is a multistep procedure involving superovulation, synchronization of donor and recipient, insemination of donor, collection, isolation, evaluation, genetic testing freezing and transfer of embryo. This is the shortest path to genetic progress on economic efficiency in large and small ruminant farms.

Conflict of interest

The authors declare no conflict of interest.

Author details

Ștefan Gregore Ciornei
Department Clinics, Biotechnology of Reproduction, Faculty of Veterinary Medicine, Iasi University of Life Sciences (IULS), Iasi, Romania

*Address all correspondence to: stefan_ciornei@yahoo.com

References

[1] Fonseca-Zangirolamo A, Kemmer Souza A, Zamparone Bergamo L, Applications of embryo transfer biotechnology in dairy cattle. SPERMOVA. 2018; 8(1): 29 – 32, DOI. 10.18548/aspe/0006.03.

[2] Colazo MG, Mapletoft RJ. A review of current timed-AI (TAI) programs for beef and dairy cattle. Canadian Veterinary Journal. 2014; 55: 772-780.

[3] Chebel RC, Demetrio DGB, Metzger J. Factors affecting success of embryo collection and transfer in large dairy herds. Theriogenology. 2008; 69(1): 98-106

[4] Vasconcelos JLM, Jardina DTG, Sá Filho OG, Aragon FL, Veras MB. Comparison of progesterone-based protocols with gonadotropin releasing hormone or estradiol benzoate for timed artificial insemination or embryo transfer in lactating dairy cows. Theriogenology. 2011; 75(6): 1153-1160.

[5] Ciornei Şt.G., Drugociu D., Rosca P, Ghinet Liliana, 2018, Polyovulatory response and in vivo embryo production at the omanian indigenous buffalo, out of breeding season in ne of Romania Lucrări Științifice, vol. 61 -IV, Medicină Veterinară, Ed. "Ion Ionescu de la Brad", Iași 286 - 291.

[6] Mircu C., coord. coordinator. Treatise on assisted reproduction (in Romanian), Ed agroprint, Timisoara, 2020, chapter 3, pages 139-168, ISBN 978-606-785-136-6

[7] Teepker G, Keller DS. Selection of sires originating from a nucleus breeding unit for use in a commercial dairy population. Can J Anim Sci 1989; 69:595-604.

[8] Seidel GE Jr. Brief introduction to whole genome selection in cattle using single nucleotide polymorphisms. Reprod Fertil Dev 2010; 22:138-144.

[9] Christensen LG. Use of embryo transfer in future cattle breeding schemes. Theriogenology 1991; 35:141-156.

[10] Pall Emoke and Cenariu M., Reproductive biotechnology, Biotehnici în reproducție (Romanian) Ed. Colorama, Cluj-Napoca, Romania, 2020.

[11] Adams GP. Control of ovarian follicular wave dynamics in cattle; Implications for synchronization and superstimulation. Theriogenology, 1994; 4:19-24.

[12] Bó GA, Adams GP, Pierson RA, et al. Exogenous control of follicular wave emergence in cattle. Theriogenology, 1995; 43:31-40.,

[13] Bó GA, Baruselli PS, Moreno D, et al. The control of follicular wave development for self-appointed embryo transfer programs in cattle. Theriogenology 2002; 57, 53-72

[14] Bó Gabriel A, Mapletoft Reuben J. Historical perspectives and recent research on superovulation in cattle. Theriogenology 2014; 81:38-48

[15] Sartori R, Bastos MR, Wiltbank MC. Factors affecting fertilisation and early embryo quality in single-and superovulated dairy cattle. Reprod Fertil Dev 2009; 22:151-158.

[16] Hasler JF. The current status and future of commercial embryo transfer in cattle. Anim Reprod Sci 2003; 79:245-264.

[17] Manisha Sethi, Nadeem Shah, Pratyush Kumar, Neeraj Bhatia, Anju Rohilla and Tushar Kumar Mohanty . Chapter - 5 Embryo Transfer Technology: Non-Surgical Collection,

Evaluation and Transfer of the Embryo in Cow, pg 83-105

[18] Selk G. Embryo transfer in cattle, 2002. (https://hdl.handle.net/11244/49938)

[19] Gengler N, Druet T. Impact of biotechnology on animal breeding and genetic progress. In Biotechnology in Animal Husbandry. Springer, Dordrecht, 2001, 33-45.

[20] Mapletoft RJ. Embryo transfer in the cow: General procedures. Revue Scientifique et Technique de l'OIE (France), 1985

[21] Seidel GE, Seidel SM. Training manual for embryo transfer in cattle. FAO animal production and health paper, 1991, 77.

[22] Dresser BL. Cryobiology, embryo transfer, and artificial insemination in ex situ animal conservation programs. Biodiversity, 1988, 296.

[23] Gordon I. Reproductive technologies in farm animals. 2nd Edition. University College Dublin. Ireland, 2017.

[24] Pickles AC. Current therapy in theriogenology, 2nd edn, edited by David A. Morrow, WB Saunders Co, Philadelphia, 1987.

[25] Mapletoft RJ, Hasler JF. Assisted reproductive technologies in cattle: a review. Revue scientifique et technique (International Office of Epizootics). 2005; 24(1):393-403

[26] Senger PL. Pathways to pregnancy and parturition. 2nd ed. Current Conceptions, Inc., Washington State University, Pullman, WA, USA, 2003

[27] Hanzen Ch., and Lourtie PVD, Follicular development in the cow. I. Morphological aspects and dynamic growth pattern, July 2000, Annales de Médecine Vétérinaire 144(4):223-+

[28] Wiltbank MC. How information of hormonal regulation of the ovary has improved understanding of timed breeding programs. In: Proceedings of the Ann Meet Soc Theriogenol 1997; 83-97.

[29] Adams GP. Control of ovarian follicular wave dynamics in mature and prepubertal cattle for synchronization and superstimulation. In: Proceeding of the XX Congress of the World Association of Buiatrics; Sydney, Australia 1998; 595-605.

[30] Mapletoft RJ, Martinez MF, Colazo MG, et al. The use of controlled internal drug release devices for the regulation of bovine reproduction. J Anim Sci 2003; 81(E. Suppl. 2): E28-E36.

[31] Ciornei S., Drugociu D., Rosca P., Ciornei Liliana. Poliovulatory response and embryo recovery rate in beef sheep in Romania, as apossibility for genetic development - A case report. Anim Reprod. 2020; 17(3), OPU/IVF and ET p.10

[32] BEAM S.W., BUTLER W.R. Energy balance, metabolic hormones and early postpartum follicular development in dairy cows fed prilled lipid. J.Dairy Sci., 1998, 81, 121-131.

[33] Verma-Kumar S, Srinivas SV, Muraly P, et al. 2004, Cloning of buffalo (Bubalus bubalis) prostaglandin F2a receptor: changes in its expression and concentration in the buffalo cow corpus luteum. Reprod; 127:705-715.

[34] Ciornei Ş. G, Drugociu D., Rosca P, Ghinet Liliana (Ciornei) 2020 - Caesarean section in uterine torsion at Buffalo (RIB) – a case study, Rev Rom Med Vet (2020) 30 | 2: 53-56, ISSN: 1220-3173; E-ISSN: 2457-7618

[35] Mapletoft Reuben J, Bó GA. Superovulation in Cattle. In: Bovine Reproduction 1st Edition, RM Hopper,

(Editor), Wiley-Blackwell Publishing, Ames, Iowa. Chapter 75; 2015; 696-702

[36] Curtis JL. Cattle embryo transfer procedure. An instruction manual for the rancher, dairyman, artificial insemination technician, animal scientist, and veterinarian. Academic Press, Inc., 1991

[37] Hafez ESE, Hafez B. Fertilization and cleavage. Reproduction in farm animals, 2000, 110-125.

[38] Valadão L, Da Silva HM, Da Silva FM. Bovine Embryonic Development to Implantation. In Embryology-Theory and Practice. Intech Open, 2018.

[39] Duran PG, Center PC. Technical Aspects of the Recovery, Handling and Transfer of Embryos. ASPAC Food & Fertilizer Technology Center, 2000.

[40] Manual of the International Embryo Transfer Society. Fourth Edition. DA Stringfellow and MD Givens (eds). Savoy, IL: IETS, 2010. Available from the International Embryo Transfer Society website.

[41] Bó Gabriel A, Mapletoft Reuben J. Evaluation and classification of bovine embryos. Anim Reprod 2013; 10:344-348

[42] Phillips PE, Jahnke MM. Embryo transfer (techniques, donors and recipients). Veterinary Clinics: Food Animal Practice. 2016; 32(2):365-385

[43] Youngquist Robert S., and Walter R. Current Therapy in Large Animal Theriogenology, 2nd Edition, Saunders. 2007

[44] Tervit HR, Cooper MW, Goold PG, Haszard GM. Non-surgical embryo transfer in cattle. Theriogenology. 1980; 13(1):63-71

[45] Hasler JF. In vitro production of cattle embryos: problems with pregnancies and parturition. Human Reproduction. 2000; 15(5):47-58.

[46] Néstor Isaías Tovío Luna, Arturo Duica Amaya and Henry Alberto Grajales Lombana. Chapter 10,Antiluteolytic Strategy for Bovine Embryo Transfer Programmes, in: http://dx.doi.org/10.5772/60425

[47] Santos J, Thatcher W, Chebel R et al. The effect of embryonic death rates in cattle on the efficacy of estrus synchronization programs. Animal Reproduction Science. 2004; 83: pp. 513-535

[48] Grajales H, Tovío N, Duica A. Fundamentos de fisiología reproductiva en la hembra bovina.1 ed. Bogota, D.C: Editorial Universidad Nacional de Colombia; 2011

[49] Binelli M, Thatcher W, Mattos R, Baruselli P. Antiluteolytic Strategies to Improve Fertility in Cattle. Theriogenology. 2001; 56: pp. 1451-1463.

[50] Silke V, Diskin M, Kenny D et al. Extent, pattern and factors associated with late embryonic loss in dairy cows. Animal Reproduction Science. 2002; 71: pp. 1-12.

[51] Tovío N, Duica A, Grajales H. Desarrollo embrionario y estrategias antiluteoliticas hormonales en programas de transplante de embriones bovinos. Revista MVZ Córdoba. 2008; 13 (1): pp. 1240-1251

[52] Silva R, Rodriguez C, Marques M et al. Efeito do eCG e do GnRH na taxa de prenhez de vacas Nelore lactantes inseminadas em tempo fixo. Acta Scientiae Veterinariae. 2004; 32 (Suplemento): pp. 221.

[53] Duica A, Tovío N, Grajales H. Factores que afectan la eficiencia reproductiva de la hembra receptora en un programa de transplante de embriones bovinos. Journal de Medicina

Veterinaria Universidad de la Salle. 2007 Jul; 14: pp. 107-124

[54] Baruselli P, Bó G, Reis E et al. Introducción de la IATF en el manejo reproductivo de rebaños de ganado de engorde en Brasil. Congreso Internacional de Reproducción Bovina Intervet. 2005; Bogotá, Colombia.

[55] Bielanski, A, and C Dubuc. 1995. In vitro fertilization of ova from cows experimentally infected with a non-cytopathic strain of bovine viral diarrhea virus. Anim. Reprod. Sci. 38:215-221. https://doi.org/10.1016/0378-4320(95)98107-F.

[56] Menchaca, A. 2020. Procedures for small ruminant embryo technology. Part 1: In vivo embryo production. Manual of the International Embryo Transfer Society. 5th ed. IETS, Champaign, IL.

[57] Thibier, M 2011. Embryo transfer: A comparative biosecurity advantage in international movements of germplasm. Rev. Sci. Tech. 30:177-188.

[58] Cottell, E, J McMorrow, B Lennon, M Fawsy, M Cafferkey, and R Harrison. 1996. Microbial contamination in an in vitro fertilization-embryo transfer system. Fertil. Steril. 66:776-780. https://doi.org/10.1016/S0015-0282(16)58635-X.

[59] Joanna Maria Gonçalves Souza-Fabjan1 and Alejo Menchaca, In: 5th Edition IETS Manual,Volume 1: 13-Supplement. Small ruminant specificity for disease control or transmission via embryos).

[60] Stringfellow D.A. (2009). – Recommendations for the sanitary handling of in vivo derived embryos. In Manual of the International Embryo Transfer Society: a procedural guide and general information for the use of embryo transfer technology emphasizing sanitary procedures, 4th Ed. IETS, Savoy, Illinois, 65-68.

[61] World Organisation for Animal Health (OIE) (2010). – Terrestrial Animal Health Code. OIE, Paris. Available at: www.oie.int/en/international-standard-setting/terrestrialcode/ access-online/ (accessed in Jun 2021).

[62] Thibier M. (2006). – Biosecurity and the various types of embryos transferred. Reprod. domestic Anim., 41, 260-267.

Chapter 4

Doppler Ultrasound in the Reproduction of Mares

Camila Silva Costa Ferreira and Rita de Cássia Lima Morais

Abstract

Doppler ultrasonographic (US) is a method that provides real-time information on vascular architecture and hemodynamic aspects of blood vessels. It can determine the presence, direction, and speed of blood flow, being subdivided into the categories of color Doppler (color flow and power flow) and pulsed Doppler. The objective of this chapter was to compile data from several studies addressing the use of US Doppler correlated with pathophysiological phenomena of equine reproduction. Initially we decided to describe the technique, advantages, and disadvantages of each Doppler mode. Then the applicability of US Doppler in mares related to equine reproduction. Thus, within this chapter, you will find the form of use and descriptions of studies carried out on vascular perfusion of the follicular dynamics, the corpus luteum, the uterine segments, which we have divided into post-insemination evaluation, endometritis diagnosis and pregnancy diagnosis. So, we hope that this chapter will expand the knowledge about US Doppler and increase the number of veterinarians who will introduce the technique into their practical routine.

Keywords: equine, diagnosis, Doppler ultrasound, reproduction

1. Introduction

In ultrasonographic (US), images of the body are obtained from the reflection or scattering of a pulsed high frequency sound beam that is sent by a mobile transducer to examine the body [1]. Each time the sound beam encounters acoustic interfaces in its path, there are changes in the density or elasticity of the medium, where the fraction of sound energy is reflected or scattered. This can happen on the walls of an organ or even along a tissue with a heterogeneous structure. The retro-scattered wave (or "echo") is detected and processed by the device, which will assign a gray scale according to the amplitude of the signal returned. Therefore, an ultrasound image corresponds to a 2D map of the tissue's acoustic reflectivity. The body can also be investigated in Doppler mode to obtain flow information, widely applied in the analysis of the circulatory system [2].

In the 1980s, the renowned researcher Dr. O.J. Ginther stated that since the introduction of transrectal palpation, ultrasound diagnosis has been the most profound technological advance in the field of research and the reproductive clinic of large animals [3]. In the late 1990s, studies were started using US Doppler to determine physiological and pathological changes in the mare's reproductive tract [4]. Over the past decades, the use of US has reached great dimensions, not only in research centers, but also commercially in livestock activities, having made great improvements in the clinical diagnosis and reproductive efficiency of large animals [5].

In this context, the objective of this chapter was to compile data from several studies addressing the use of US Doppler correlated with pathophysiological phenomena of equine reproduction.

2. Doppler ultrasonographic

US Doppler is a method that provides real-time information on vascular architecture and hemodynamic aspects of blood vessels. It can determine the presence, direction and speed of blood flow, being subdivided into the categories of color Doppler (color flow and power flor) and pulsed Doppler [6].

In color flow Doppler US (UCF) is considered the classic examination of the color mode within US Doppler, which allows non-invasive assessment of the presence, direction, speed, and quality of blood flow. Two distinct colors are used, usually variations of red tones for positive flows and blue for negative flows representing the vascular blood perfusion of a structure, where the colored pixels indicate the direction of the blood cells in relation to the transducer [7–9].

In power flow Doppler US (UPF) is a variation of the UCF and has been used as the test of choice to evaluate the vascularization of the uterus and ovaries of mares and components of the testicular bag of the stallions [7, 9–15]. It is observed with this technique an increase of the sensitivity of display of the blood flow inside the tissue of 3 to 5 times, in comparison with UCF. The higher sensitivity allows the evaluation of vessels with a small diameter or slow flow that does not appear in a conventional color flow because of incompatible speed ranges and Doppler angles. [16]. Blood flow is generally characterized homogeneously by orange tones and the speed of flow is indicated in both modes by the intensity of the tones, the lighter the color the faster the flow and the darker the color the slower the flow [17]. In this way, the advantages of UPF are greater sensitivity to weak flow; effects of the Doppler angle on the Doppler frequency are ignored; and aliasing does not affect the display of colors [9].

The extent of vascularization of the color US Doppler can be estimated subjectively through the percentage of pixels (colored signs) of a tissue or objectively by counting the colored pixels via software [10].

In spectral Doppler US mode, the artery blood flow wanted is found and then the cursor ("gate") is positioned in the lumen of the artery to evaluate the sample volume [18]. The volume evaluations of the samples are reflected in graphs, called spectrum, which represent the speed of the blood flow of the artery in question for several times within the cardiac cycle or of an individual arterial pulse [19]. The brightness of the spectral trace, represented in the gray scale, is also used to represent the amplitude of each frequency component, indicating the amount of blood cells that pass at a particular speed [6]. Changes in the evaluated vessel may suggest changes in the spectral tracing and this fact may be indicative of physiological, systemic changes or of some disease present at the site [20].

In the quantitative analysis of the spectral Doppler US tracing, most devices have automatic configuration to automatically calculate the average of the displacement frequency or speed. The maximum point reached in the spectrum is called the systolic peak velocity (PSV) and the minimum point in the wave morphology is the value of the final diastolic velocity (EDV). The medium flow can be calculated by multiplying the average speed by the vessel area [9, 20].

Hemodynamic indices, such as systole-diastole ratio, resistivity index (RI) and pulsatility index (PI), allow the comparison of flow during systole and diastole. RI and PI have a negative correlation with the vascular perfusion of the tissue irrigated by the artery in question, that is, the lower the RI and PI, the greater the vascular perfusion in the tissue supplied by that vessel [10]. The changes in these indexes help in the identification of stenosis, thrombosis and changes in vascular resistivity and parenchyma dysfunctions, or in the characterization of disease malignancy [6].

3. Use in mares reproduction

3.1 Follicular dinamic

The UCF has the potential to predict the follicular state (ovulatory or anovulatory future) of the dominant follicles during the transition period. During the anovulatory transition season, vascular changes in the walls of the future anovulatory follicle and of a future dominant ovulatory follicle were studied from 25 mm in 7 days after the follicle was 30 mm. The blood flow area was smaller for anovulatory follicles of dominant size than for ovulatory follicles at the time when blood flow determinations started at 25 mm [21, 22]. There is a hypothesis for anovulation that involves hormones and follicular angiogenesis during the transition period [23]. In this regard, pre-ovulatory vascular changes were compared between the first and the next ovulation of the year in 40 pony mares for 6 days before ovulation [24]. Although the area of blood flow from the follicle increased the day of ovulation in both groups, the results demonstrated that follicle vascularization and LH peak were attenuated before the first ovulation of the year, with no indication that estradiol was involved in the differences between the first and the last ovulation.

Regarding the blood flow of the preovulatory follicle, recent studies have shown a daily increase in vascularization of the dominant follicle wall as it matures and approaches the day of ovulation [22, 24–26]. However, on the day of ovulation, a few hours before ovulation, an abrupt decrease in blood perfusion was detected in the wall of the preovulatory follicle [9, 25].

Vascular perfusion is related to the peripheral area of the follicle and its diameter [19]. Therefore, research shows that a higher pregnancy rate is associated with a greater blood flow in the preovulatory follicle (POF). Because a higher percentage of Doppler signals was observed in the follicular wall and a reduction in the Doppler indices of RI and PI in the ovarian vessels in mares that impregnated in comparison to mares that did not impregnate [27].

Another study demonstrated that larger, well-vascularized POF produce larger corpus luteum (CL), with more blood flow and a higher systemic concentration of progesterone (P4), which can lead to a better uterine environment for the establishment of pregnancy. In addition, the repetition of the POF diameter value in individuals during spring and autumn may be estimate the best breeding time during the transition period. The lower blood flow of CL observed during the last estrous cycle of the reproductive season is another important finding of the study, which may clarify the luteal insufficiency of the transition period [28].

The morphology and vascularization of anovulatory hemorrhagic follicles (HAF) in mares and the endocrinology immediately prior to the formation of HAF were studied in control and HAF groups [29]. The day of ovulation and the first day of HAF formation, as indicated by the turbidity of the follicular fluid, were defined as Day 0. The frequency of discrete ultrasound indicators in imminent ovulation gray scale and the diameter of the follicle on Day −1 did not differed between the groups with future ovulation and HAF. However, the circumference of the follicle wall of future HAF had more signs of color Doppler than in control mares [25].

3.2 Corpus luteum

The UCF can be effective in identifying mares that fail or regress CL, before any decrease in P4 circulation, decrease in CL area, echogenicity changes in B-mode image or when uterine/cervical tone becomes apparent [30].

A daily assessment of CL between the day of ovulation and the eighth day after ovulation was performed by Romano et al. [15]. The authors observed that the CL area was weakly correlated with luteal vascular perfusion and plasma P4 concentrations. However, a positive correlation was observed between luteal vascular perfusion and plasma P4 concentrations. Furthermore, the number of colored pixels and the total pixel intensity were positively correlated with vascular perfusion and the plasma P4 concentration.

From a daily analysis, via UPF of the CL, a transient increase in the total luteal area was observed during the first days after ovulation, demonstrating that old mares had a greater luteal area than young mares between D2 and D8 and in D18-D19 ($p < 0.05$). However, old mares have a late increase in luteal vascularization during the first gestational days ($p < 0.05$). However, the CL of young and old mares showed similar and constantly high vascularization from D14. It has also been observed that the progressive increase in plasma P4 concentrations observed up to D8 was followed by a gradual decrease until to intermediate levels of P4. Thus, concluding that the newly formed CL of old mares underwent a compensatory structural remodeling to guarantee the local blood supply and the continuous output of P4 during early pregnancy to maintain it without suffering from the age effect [11].

To check if there could be a difference between the vascularization of the CL between the possible days for embryo transfer in the recipient mares, it was observed that the CL classified as adequate for the embryo transfer procedure by US mode B were also those classified with the better vascularization by UPF maintaining the pattern of proportional growth of the size of the CL with the concentration of P4 [31].

De Vasconcelos Azevedo et al. [32] evaluated the vascularization and function of CL of recipient mares of the Mangalarga Marchador embryo at the time of embryo transfer through UCF. As a result, they noticed that the mares that were pregnant showed a correlation with the increase in the vascularization of the CL, as well subjective methods as objective methods, and the plasma concentration of P4.

3.3 Uterine evaluation

The hemodynamic evaluation of the uterus can be done by spectral data collected from the uterine arteries and their ramifications [33], or by subjective or objective assessments of the blood flow of the endometrium that provide data regarding local and specific changes in the evaluated area [8, 34].

The first investigations relating US Doppler to equine uterine physiology were carried out by Stolla and Bollwein [35], Mayer et al. [36] and Bollwein et al. [33], who obtained a cyclic pattern in uterine blood flow. Bollwein et al. [33] performed the measurement of the RI of the uterine artery on different days of the equine estrous cycle. They are verifying that there was no difference between the values of the left and right arteries. Not showing a correlation of the values found in the flow with the presence or absence of follicles or luteum corpus ipsilateral to the evaluated horn, indicating that the circulation is distributed equally to the two horns of the uterus. The authors also found that the average RI for all evaluated days was higher when observed in multiparous mares than when observed in nulliparous mares. This work also found significantly higher RI values on days 0 and 10 of the estrous cycle compared to days 5, 15 and 20 (considering D0 the day of ovulation). The authors correlated the lower RI value observed on day 5 to the possibility of an increase in blood supply to the uterus in this initial luteal phase due to the moment of embryo entry into the uterus.

Subsequent studies by Bollwein et al. [37], confirmed these statements, in addition to noting that the lowest uterine PI values were record during the initial diestrus, in a stage, when they found a peak in uterine tonus in mares. They speculated, therefore, that the tone and contractility of the uterus do not seem to be regulated by the uterine blood supply in cyclic mares.

Ferreira et al. [31] evaluated the flow of the uterine artery of recipient's mares on the day of embryo transfer. It was observed that in animals where the RI of the dorsal branches of the uterine arteries close to 1.0 are proportional to mares with greater vascularization of the corpus luteum and a high plasma P4 concentration. Being able to use these indices to select mares with greater aptitude for the development of the embryo when comparing two similar mares clinically by conventional ultrasound exams.

Other studies correlating the uterine flow of normal mares with subfertility characterized by biopsy endometrial were performed by Stolla and Bollwein [35] and Blaich et al. [38]. The authors show that in all mares there was an increase in vascular resistance in the pre-ovulatory phase and in 8 mares the peak occurred 8 days after ovulation. The blood flow impedance decreased from D1 after ovulation until it reached its lowest level during a luteal phase in all mares. They also observed that there was a gradual increase in vascular resistance in the initial follicular phase and that cyclical changes occurred in all mares with varying amplitudes. As they compared the flows according to the endometrial classifications, they observed that, for all mares with histological category IIb and III, there was an increase in the PI and average RI throughout the estrous cycle, when compared with the mares in the group that had normal fertility.

3.3.1 Uterine evaluation after artificial insemination

Although adequate blood flow is essential for the normal functioning of the reproductive system [9], there are few in vivo studies that describe the uterine hemodynamics of non-pregnant mares [34, 39].

Changes in blood flow velocity verified by spectral Doppler US of the dorsal uterine arteries after reproduction, suggest that there is an increase in endometrial blood flow during semen transport and uterine clearance [40]. However, currently, the evaluation of the Doppler indices of the mesometrial arteries and the vascular perfusion of the uterine tissue during the pre- and post-reproductive periods and in mares with endometritis have been little researched.

Bollwein et al. [41] also evaluated, during three estrous cycles, the vascular perfusion of the uterine artery using the mean maximum velocity spectral Doppler index (MVM). The authors related the quantification of flow to the effect of the infusion of semen extender, seminal plasma, or pure semen. In response to the dilution infusion, no effect on uterine blood flow was observed. Controversial results have been reported after one hour of infusion of seminal plasma or pure semen, where an increase in VMM values was observed in both uterine arteries. Therefore, the authors concluded that the increase in endometrial perfusion in these groups may be associated with inflammation and vasodilator components present in the seminal plasma.

Ferreira et al. [8] observed a transient increase in uterine vascular perfusion without mesometrial changes in the PI during the first 8 hours after artificial insemination. However, in the research by Ferreira et al. [42, 43], where PI Doppler measurements of the mesometrial arteries and UPF of the organ in relation to the semen effect were used, changes in the blood flow velocity of the uterine arteries were observed only in the first hour after the infusion of crude semen.

Ferreira et al. [42, 43] also evaluated uterine vascular perfusion before and after artificial insemination correlating with the age of the mares and the presence/absence of endometrial degenerative processes. There were no differences in perfusion between the horns, however, they showed in the organ, an early and transient increase in the blood flow of the uterus in response to artificial insemination in all mares. However, the increase in mesometrial arterial resistance was strongly associated with severe endometrial degenerative changes after AI, regardless of age.

In horses, the AI procedure can be performed by depositing the semen in the body of the uterus or at the apex of the uterus horn to use a reduced inseminating dose. In order to assess the uterine inflammatory response to the different semen deposition sites, Araújo [44] analyzed the uterine perfusion by means of spectral Doppler US of the dorsal uterine arteries and with subjective UCF in the uterus. The values of RI and PI identified after AI were similar on the contralateral and ipsilateral sides of the mesometrial and dorsal uterine arteries. Regarding the evaluation time, a difference was observed where, both the RI and the PI, were lower in the moment before ovulation and AI and higher values were recorded in 24 h after. As for the analysis of organ perfusion by score, this did not show any difference within each experimental group regarding the evaluation time. Regarding the place of insemination, the color scores showed significant differences between the two experimental groups, with a predominance of scores 1.0 and 2.0 in the group of mares inseminated at the apex of the uterine horn, and the score 3.0 in the group inseminated in the body of the uterus.

3.3.2 Endometritis' diagnosis

A research to measure uterine blood perfusion using UCF (subjective and objective) was carried out by Sá et al. [45] to investigate changes in vascularization of uterine segments of mares after intrauterine inoculation of E. coli. The authors found significant differences in the blood flow evaluated before inoculation of the bacteria (M0) and 24 h after inoculation (M1), where blood perfusion in M1 showed almost twice the M0 in the three uterine segments evaluated, but no significant differences were found in evaluations carried out between the follow-ups. In four animals evaluated, endometrial cytology and mode B ultrasound examinations were not sufficient to detect uterine infection, however, through UCF, it was possible to verify a significant increase in uterine vascularization in these mares compared to the values before inoculation.

In a second experiment carried out by Sá et al. [45], the evaluation by the UCF was performed 24 h after the intrauterine infusion of E. coli. Half of the mares with uterus inoculated by bacteria were subjected to a treatment consisting of a uterine lavage using 50 mL of the Fitoclean® phytotherapy solution (Organnact Saúde Animal, Brazil) diluted in 950 mL of Ringer with Lactate and subsequent intrauterine infusion with 40 mL phytotherapy solution Fitoclean® diluted in 60 mL of Ringer with Lactate. In the control group, an intrauterine wash was performed with Ringer's serum with pure lactate, with subsequent infusion of 100 mL of the same substance. UCF was performed before treatment (A), 24 h (B) and 48 h (C) after treatment. The authors reported that uterine perfusion was greater at time A than at time B and C, but the decrease in uterine bacterial load was not verified. The authors assume that the decrease in vascularization in the post-treatment groups can be attributed to the vasoconstriction caused by the reaction of components with anti-inflammatory properties present in the product, which may have caused the decrease in local perfusion even with the presence of microorganisms.

A third research was carried out in the study by Sá et al. [45]. The mares in which pathogenic agents were identified in the samples collected 10 days after the end of the second mentioned experiment, were subjected to antibiotic therapy by intrauterine infusion of 100 mL Gentamicin (Gentrin® Uterine Infusion, Ourofino Saúde Animal, São Paulo, Brazil) for three days, according to the sensitivity shown in the antibiogram. UCF was performed seven days after treatment, comparing the tests performed before E. coli pre-inoculation, post-treatment with Fitoclean® and one week after antibiotic therapy (M0, M1 and M2, respectively). The authors observed a reduction in vascularization in the group treated with Fitoclean® after antibiotic therapy, however, blood perfusion in this group was still greater than M0, even with antibiotics being able to eliminate the bacteria present. Based on the results obtained in this study, it was possible to identify acute endometritis through the UCF, but the vascular perfusion identified did not correlate with the uterine laboratory tests performed.

Abdelnaby et al. [46] performed spectral Doppler ultrasonography of the dorsal uterine arteries and UCF of the uterus of mares with and without endometritis, correlating the data with the impact of the pathology on the oxidative and hormonal state. The results revealed a significant increase in the metabolites of estradiol, malondialdehyde and nitric oxide associated with a significant decrease in progesterone and total antioxidant capacity in the endometritis group. The uterine blood flow analyzed by the UCF showed a significant increase in the endometritis group, while the spectral mode showed a significant increase in the PSV and TAMV indices and in the blood flow area rate accompanied by a significant decrease in the PI and RI Doppler indices. In addition, the elevated uterine blood perfusion was correlated with the accumulation of fluid inside the uterus, with a marked difference between the uterine horns in relation to the size of the UCF staining area, which may be due to the marked increase in fluid accumulation in the right in endometritis group.

Morais [13] evaluated, through UPF, the intensity of colored pixels (IPC) of the uterine segments (body and horns) during estrus before and after uterine treatment with DMSO. From the evaluations it was noted that mares with negative cytology (despite positive culture test) had less blood flow than mares with positive cytology. IPC was reduced in mares that became pregnant. In the mares that remained empty, the CPI remained high. Thus, UPF can be used as an auxiliary diagnostic method in some cases of equine endometritis. In pregnant mares, blood flow in the endometrium and RI decreased. Empty mares blood flow in the endometrium and RI remained high or increased.

3.4 Doppler ultrasonography in pregnancy diagnosis

Changes in the hemodynamics of the equine reproductive tract during early and late pregnancy have already been described several times in the literature. Bollwein et al. [40] to compare the vascularization of the right and left uterine horns of cyclical mares and in initial pregnancies, observed that on days 11 and 15 to 29 of gestation, mean values of blood flow velocity were higher and RI lower in both horns uterine of pregnant mares compared to empty mares.

Subsequently, Bollwein et al. [47] investigated the blood flow in both uterine arteries (ipsilateral and contralateral to the fetus) every four weeks, from the second week until the moment of delivery. A highly significant regression was observed in the RI averages according to the week of gestation, reaching, at the end of the evaluations, values lower than half of those initially recorded (0.89 ± 0:01 to

0.39 ± 0:03). The volume of blood flow in the ipsilateral and contralateral uterine horns increased significantly according to the week from the middle of gestation.

In other research, it was shown that the transient changes in vascular perfusion accompany the mobility and the fixation of the embryonic vesicle [34]. Pregnant and non-pregnant mares have similar and low endometrial vascularity in the first eight days after ovulation. However, from D11 there was a gradual increase in the volume of blood flow in both uterine horns during the embryonic mobility phase and a higher speed in blood flow in the uterine horn of fixation of the conceptus, in relation to the opposite side.

Silva and Ginther [29] observed, through UCF, an early vascular indicator of the future position of the embryo itself, which consisted of a colored point in the image of the endometrium close to the wall of the embryonic pole. The early indicator was detected in each mare 0.5 ± 0.1 days after fixation and 2.5 ± 0.2 days before the first visible embryo detection. The author also reports that by US Doppler we can monitor the inadequate early orientation of this embryo. Studies have shown that this type of problem is correlated with a flabby uterus and a defective embryonic dorsal invasion of the endometrium. However, the asymmetric increase in the allantoic sac can spontaneously correct the disorientation, so that the orientation for the formation of the umbilical cord is in a normal position around 12 o'clock.

Using spectral-mode US Doppler, Chen and Stolla [48] developed the uterine index (UI) to predict embryonic death in mares. The calculation used by the authors considers the following formula: UI = (RI-p - RI-np) x100, where RI-p is the RI of the uterine artery on the side of the pregnant uterine horn and RI-np is the RI of the artery uterine horn of the non-pregnant uterine horn. According to the authors, UI less than five is indicative of embryonic death evident in the next 24 hours, while mares with UI more than 10 did not present any apparent hemodynamic disorder.

Ferreira et al. [12] observed that the uterine doppler indices (RI and PI) of pregnant mares decreased progressively. Unlike other studies, this research also observed, through measurements of mesometrial RI and PI, an increase in vascular perfusion between D3 and D6 post-OV. This study was one of the triggers that helped future inquiries regarding the use of the technique as an early pregnancy diagnosis and determination of the ideal moment for embryo collection, especially when aimed at cryopreservation.

Ousey et al. [49] evaluated the blood flow of the uterine artery and other Doppler indices during pregnancy to compare placental and fetal development in young and elderly mares. However, no difference was found in the evaluated indexes. Thus, the authors stated that the similarity in the Doppler indices between the groups of elderly and young mares reflects the absence of severe pathological changes in the endometrial vascularization and glandular tissue found in the elderly mares used. However, employing other types of blood perfusion measurements, it has already been observed that mares with diffuse endometrial degeneration had reduced uterine vascular perfusion when compared to pregnant mares with unchanged endometrium. The hypothesis was then raised that severe angiosis can reduce the capacity of the vessels to adapt to the varied demands of uterine circulation [50]. In addition, endometrial pathological changes have been strongly associated with degeneration of uterine vessels in mares [51].

Ferreira et al. [42], used in their research, in addition to the analysis of uterine perfusion with UPF, a spectral analysis of the insertion of uterine arteries in the mesometrium, which is also reported to be an efficient method for the objective examination of uterine blood flow during the first 20 days gestation. With this methodology, the authors detected early and transient increases in uterine blood flow in pregnant mares, regardless of age and presence of endometrial changes. The increase in uterine blood flow in the initial pregnancy described by this and other authors, may be caused by the effects of vasoactive factors sought by pre-implanted embryos, as described in horses [34].

Another analysis carried out in the study by Ferreira et al. [42], was an observation of the negative effect of age and endometrial degeneration without uterine blood flow from mares during early pregnancy. The authors associated this apparent inability of the aged uterus to respond to vasoactive factors derived from the embryo with the progressive changes in the architecture of the uterine vascular network observed in older mares. In addition, it has already been added that age-related degenerative changes in the endometrium can affect the development of placental microcotyledons and their associated blood flow [49].

With UCF it is also possible to determine fetal equine sex between 90 and 180 days of gestation by observing fetal gonads [52]. This assessment was later confirmed by Pricking et al. [53], who obtained high rates of sex determination accuracy when associated with Us mode B, color Doppler and 3D transabdominal tomography ultrasound techniques. In addition, the application of this new 3D imaging ultrasound technology has enabled the diagnosis of gender in 18 cases in which B-mode and Doppler ultrasonography have shown dubious results.

In relation to the gestation of hybrids, pregnant donkey showed greater uterine vascularization in the uterine horn contralateral to that of embryonic fixation. And that older mares showed less blood flow in the uterine artery, as they had higher PI and RI [54]. Another difference observed was comparing the PI of the fetal umbilical cord artery, which showed a statistical difference in pregnant with donkey semen when compared to pregnant with stallion semen [55].

Nieto-Olmedo et al. [14] detected differences in uterine vascular perfusion between pregnant and non-pregnant mares early between days 7 and 8 post-ovulation, demonstrating that the UPF associated with computerized analyzes is an effective method for the early diagnosis of pregnancy. The authors observed that the area of vascular perfusion of the uterus (mm^2) and pixel intensity increases in pregnant mares compared to mares without embryonic recovery. The technique can be used in routine clinical practice to maximize the embryo recovery rates of donor mares and to predict the diagnosis of pregnancy before embryo collection.

Relationship of uterine vascular perfusion associated with involution of the postpartum uterus was mentioned in the study by Lemes et al. [56] using the UCF mode of uterine segments. The authors observed that vascular perfusion increased in endometrial and mesometrial tissues in the first 2 to 4 days after delivery, followed by a progressive decrease until the second week postpartum. The profile of vascular perfusion in the uterus described after the first postpartum ovulation is similar to that observed during the estrous cycles and initial gestation observed previously, indicating a rapid return of the uterus to the pre-gestational uterine characteristics in the mares. In such cases, it has been reported that the rapid reduction in uterine diameter, absence of intrauterine fluid and vascular decrease in the layers of the uterus, may indicate a favorable uterine environment for the development of the embryo in the foal's heat.

Author details

Camila Silva Costa Ferreira* and Rita de Cássia Lima Morais
Federal Rural University of Rio de Janeiro, Seropédica, RJ, Brazil

*Address all correspondence to: vetcamilaferreira@gmail.com

References

[1] BUSHBERG, J.T.; SEIBERT, J. A; LEIDHOLDT, JR. E.M.; BOONE, J.M. The essential physics of medical imaging. Filadélfa: Lippincott Williams & Wilkins; 2002.

[2] PAPALÉO, R, M.; DE SOUZA, D.S. Ultrassonografia: princípios físicos e controle da qualidade. Revista Brasileira de Física Médica. v.13(1), p.14-23, 2019.

[3] GINTHER, O.J. Ultrasonic imaging and reproductive event in the mare. Cross Plains, WI,Equiservices, Inc. p.1-12, 1986.

[4] EL-SHAHAT, K. H.; EL-MAATY, A.A.; HELMY, M.; EL BAGHDADY, Y. Power and colour Doppler ultrasonography for evaluation of the ovarian and uterine haemodynamics of infertile mares. Bulgarian Journal of Veterinary Medicine, 2019.

[5] PUGLIESI, G.; REZENDE, R. G.; SILVA, J. C. B.; LOPES, E.; NISHIMURA, T. K.; BARUSELLI, P.S.; MADUREIRA, E. H.; BINELLI1, M. Uso da ultrassonografia Doppler em programas de IATF e TETF em bovinos. Revista Brasileira de Reprodução Animal, v.41, n.1, p.140-150, 2017.

[6] CARVALHO, C. F.; CHAMMAS, M. C.; CERRI, G. G. Princípios físicos do Doppler em ultrassonografia: revisão bibliográfica. Ciência Rural, Santa Maria, vol. 38, n. 3, p. 872-879, 2008.

[7] DASCANIO, J.J.; MCCUE, P.M. Equine reproductive procedures. Aspen: Wiley Blackwell, 560p, cap. 17, p. 58 - 62, 2014.

[8] FERREIRA, J.C.; IGNÁCIO, F.S.; MEIRA, M. Spectral and Power-flow doppler imaging evaluation of the equine uterus during the post-breeding. 12th Congress of WEVA, 2011.

[9] GINTHER, O.J. Ultrasonic imaging and animal reproduction: Color-Doppler ultrasonography – Book 4. Madison: Equiservices Publishing, 258p, 2007.

[10] BOLLWEIN, H.; HEPPELMANN, M.; LÜTTGENAU, J. Ultrasonographic Doppler use for female reproduction management. Veterinary Clinics: Food Animal Practice, v. 32, n. 1, p. 149-164, 2016.

[11] FERREIRA, J. C.; BOAKARI, Y. L.; ROCHA, N. S.; IGNACIO, F. S.; COSTA, G. B.; MEIRA, C. Luteal vascularity and embryo dynamics in mares during early gestation: Effect of age and endometrial degeneration. Reproduction in domestic animals, v. 54, n. 3, p. 571-579, 2019.

[12] FERREIRA, J.C.; IGNÁCIO, F.S.; MEIRA, C. Uterine vascular perfusion and spectral-Doppler measurements during early gestation in mares: New concepts of evaluation. Animal Reproduction Science, v. 121, p. 281-283, 2010.

[13] MORAIS, R.C.L. Efeito da infusão intrauterina de dimetilsulfóxido sobre a vascularização e o perfil microbiológico uterino em éguas com endometrite, 131F (Tese), Universidade Federal Rural do Rio de Janeiro, 2021.

[14] NIETO-OLMEDO, P.; MARTÍN-CANO, F.E.; GAITSKELL-PHILLIPS, G.; ORTIZ-RODRÍGUEZ, J.M.; PEÑA, F.J.; ORTEGA-FERRUSOLA, C. Power Doppler can detect the presence of 7-8-day conceptuses prior to flushing in an equine embryo transfer program. Theriogenology, v. 145, 2020.

[15] ROMANO, R. M.; FERREIRA, J. C.; CANESIN, H. S.; BOAKARI, Y. L.; IGNACIO, F. S.; NOVAES FILHO, L. F. THOMPSON JR., D. L.; MEIRA, C. Characterization of luteal blood flow and secretion of progesterone in mares

treated with human chorionic gonadotropin for ovulation induction or during early diestrus. Journal of Equine Veterinary Science, v. 35, n. 7, p. 591-597, 2015.

[16] GINTHER, O.J.; GASTAL, E.L.; GASTAL, M.O.; BEG, M.A. Seasonal influence on equine follicle dynamics. Animal Reproduction, v.1, p. 31-44, 2004.

[17] FOGAÇA, JESSICA LEITE et al. Modalidade Doopler na Ultrassonografia. 5° Jornada Científica e Tecnológica da FATEC de Botucatu, São Paulo, Brasil. 2016.

[18] GINTHER, O.J; Utt, M.D. Doppler ultrasound in equine reproduction: principles, techniques, and potential. Journal of Equine Veterinary Science, v. 24, n. 12, p. 516-526, 2004.

[19] GINTHER, O. J. How ultrasound technologies have expanded and revolutionized research in reproduction in large animals. Theriogenology, v. 81, n. 1, p. 112-125, 2014.

[20] CARVALHO, C. F. Ultrassonografia Doppler em pequenos animais. São Paulo: Roca, 274p, 2009.

[21] ACOSTA, T.J.; BEG, M.A.; GINTHER. O.J. Aberrant blood flow area and plasma gonadotropin concentrations during the development of dominant-sized transitional anovulatory follicles in mares. Biol Reprod, v.71, p.637-642, 2004.

[22] PALMER, E.; CHAVATTE-PALMER, P.; VERDONCK, E. Field trial of Doppler ultrasonography of the preovulatory follicle in the mare. Anim Reprod Sci, v.94, p.182-183, 2006. Abstract

[23] GINTHER, O.J.; GASTAL, E.L.; GASTAL, M.O.; BEG, M.A. Seasonal influence on equine follicle dynamics. Anim Reprod. 1, p.31-44, 2004b.

[24] GASTAL, E.L.; GASTAL, M.O.; DONADEU, F.X.; ACOSTA, T.J.; BEG, M.A.; GINTHER, O.J. Temporal relationships among LH, estradiol, and follicle vascularization preceding the first compared with later ovulations during the year in mares. Anim Reprod, v.102, p.314-321, 2007.

[25] GASTAL, E.L.; GASTAL, M.O.; GINTHER, O.J. Relationships of changes in B-mode echotexture and colour-Doppler signals in the wall of the preovulatory follicle to changes in systemic oestradiol concentrations and the effects of human chorionic gonadotrophin in mares. Reproduction, v.131, p.699-709, 2006

[26] GINTHER, O.J.; GASTAL, M.O.; GASTAL, E.L.; JACOB, J.C.; BEG, M.A. Age-related dynamics of follicles and hormones during an induced ovulatory follicular wave in mares. Theriogenology, v.71, p.780-788, 2009.

[27] SILVA, L. A.; GASTAL, E. L.; GASTAL, M. O.; BEG, M. A.; GINTHER, O. J. Relationship between vascularity of the preovulatory follicle and establishment of pregnancy in mares. Anim Reprod, v. 3, n. 339, p. e46, 2006.

[28] ISHAK, G. M.; BASHIR, S. T.; GASTAL, M. O.; GASTAL, E. L. Pre-ovulatory follicle affects corpus luteum diameter, blood flow, and progesterone production in mares. Animal reproduction science, v. 187, p. 1-12, 2017.

[29] GINTHER, O.J.; GASTAL, E.L.; GASTAL, M.O.; BEG, M.A. Conversion of a viable preovulatory follicle into a hemorrhagic anovulatory follicle in mares. Anim Reprod, v.3, p.29-40, 2006.

[30] BROGAN, P. T.; HENNING, H.; STOUT, T. A. E.; RUIJTER-VILLANI, M. de. Relationship between colour flow Doppler sonographic assessment of corpus luteum activity and progesterone

concentrations in mares after embryo transfer. Animal reproduction science, v. 166, p. 22-27, 2016.

[31] FERREIRA, C.S.C.; MORAIS. R.C.L.; ANDRADE A.B.P.; BALARO, M.F.A.; RIBAS, J.A.S.; GOMES, G.M.; PINNA, A.E. Spectral Doppler Ultrasound in selecting an equine embryo receiver. Reproduction in Domestic Animals. 2020

[32] DE VASCONCELOS AZEVEDO, Marlon et al. Evaluation of Corpus Luteum Vascularization in Recipient Mares by Using Color Doppler Ultrasound. Acta Scientiae Veterinariae, v. 49, 2021.

[33] BOLLWEIN, H.; MAIERL, J.; MAYER, R.; STOLLA, R. Transrectal color Doppler sonography of the A. uterina in cyclic mares. Theriogenology, v.49, p.1483-1488, 1998.

[34] SILVA, L.A.; GASTAL, E.L.; BEG, M.A.; GINTHER, O.J. Changes in vascular perfusion of the endometrium in association with changes in location of the embryonic vesicle in mares. Biology of Reproduction, v.72, p.755-761, 2005.

[35] STOLLA, R; BOLLWEIN, H. Color doppler sonography of the uterine artery in subfertile mares. Pferdeheilk. v. 3, p.547. 1997.

[36] MAYER, R., BOLLWEIN, H. AND R. STOLLA. Uterine blood flow during the estrous and early pregnancy in mares studied by transrectal colour Doppler sonography. Reprod. Dom. Anim. v.33, p.71, 1998.

[37] BOLLWEIN, H.; MAYER, R.; WEBER, F.; STOLLA, R. Luteal blood flow during the estrous cycle in mares. Theriogenology, v. 65, p. 2043-2051, 2002.

[38] BLAICH, U.; PETZOLD, S.; BARTMANN, C.P.; LORBER, K.; SCHOON, H.-A.; RAILA, G.; BOLLWEIN, H. Doppler ultrasonographic studies of the uterine blood flow in mares. Pferdeheilkunde 15, 547 - 550, 1999.

[39] FERREIRA, J.C.; GASTAL, E.L.; GINTHER, O.J. Uterine blood flow and perfusion in mares with uterine cysts: effect of the size of the cystic area and age. Reproduction, v. 135, p. 541 – 550, 2008.

[40] BOLLWEIN, H.; MAYER, R.; STOLLA, R. Transrectal Doppler sonography of uterine blood flow during early pregnancy in mares, Theriogenology, V. 60, Ed. 4, p. 597-605, 2003a.

[41] BOLLWEIN, H.; SOWADE, C.; STOLLA, R. The effect of semen extender, seminal plasma and raw semen on uterine and ovarian blood flow in mares, Theriogenology, V. 60, Ed. 4, p. 607-616, 2003b.

[42] FERREIRA, J.C.; CANESIN, H.S.; IGNÁCIO, F.S.; ROCHA, N.S.; PINTO, C.R.; MEIRA, C. Effect of age and endometrial degenerative changes on uterine blood flow during early gestation in mares. Theriogenology, 2015a.

[43] FERREIRA, J.C.; IGNÁCIO, F.S.; ROCHA, N.S.; THOMPSON, D.L.; PINTO, C.R.; MEIRA, C. Real-time characterization of the uterine blood flow in mares before and after artificial insemination, Animal Reproduction Science, V.160, p. 90-96, 2015b.

[44] ARAÚJO, A. G. Avaliação da resposta inflamatória em útero de éguas após diferentes técnicas de inseminação artificial utilizando sêmen asinino. Niterói. 2014. 71f. Dissertação (Mestrado) – Universidade Federal Fluminense, 2014

[45] Sá, M.A.F.; LIMA, T.F.G.; MORAIS, R.C.L.; DUTRA, G.A. JESUS, V.L.T.;

JACOB, J.C.F. Acute endometritis induced by Escherichia coli in mares evaluated through color doppler ultrasonography, Arquivo Brasileiro de Medicina Veterinária e Zootecnia, v.72, n.5, p.1586-1598, 2020.

[46] ABDELNABY, E. A.; EMAM, I.A.; SALEM, N.Y.; RAMADAN, E.S.; KHATTAB, M.S.; FARGHALI, H.A.; ABD EL KADER, N.A. Uterine hemodynamic patterns, oxidative stress, and chromoendoscopy in mares with endometritis, Theriogenology, 2020.

[47] BOLLWEIN, H.; WEBER, F.; WOSCHÉE, I.; STOLLA, R. Transrectal Doppler sonography of uterine and umbilical blood flow during pregnancy in mares, Theriogenology, V. 61, Ed. 2-3, p.499-509, 2004.

[48] CHEN, Y.H.; STOLLA, R. Using uterine index to diagnose embryonic death in mares. Journal of Equine Veterinary Science, v.26, p.219-224, 2006.

[49] OUSEY, J.C.; KÖLLING, M.; NEWTON, R. WRIGHT, M.; ALLEN, W.R. Uterine haemodynamics in young and aged pregnant mares measured using Doppler ultrasonography. Equine Veterinary Journal Supplements, v.41, p.15-21, 2012

[50] GRÜNINGER, B.; SCHOON, H.A.; SCHOON, D.; MENGER, S.; KLUG, E. Incidence and morphology of endometrial angiopathies in mares in relationship to age and parity. Journal of Comparative Pathology, v.119, p. 293-309, 1998.

[51] ESTELLER-VICO, A.; LIU, I. K.; VAUGHAN, B.; STEFFEY, E.P.; BROSNAN, R.J. Effects of vascular elastosis on uterine blood flow and perfusion in anesthetized mares. Theriogenology 2015.

[52] RESENDE, H. L.; CARMO, M. T.; RAMIRES NETO, C.; ALVARENGA, M. A. Determination of equine fetal sex by Doppler ultrasonography of the gonads. Equine veterinary journal, v. 46, n. 6, p. 756-758, 2014.

[53] PRICKING, S.; SPILKER, K.; MARTINSSON, G.; RAU1, J.; TÖNIßEN1, A.; BOLLWEIN, H.; SIEME, H. Equine fetal gender determination in mid- and advancedgestation by transabdominal approach – comparative study using 2D B-Mode ultrasound, Doppler sonography, 3D B-Mode and following tomographic ultrasound imaging. Pferdeheilkunde, v.35(1), p.11-19, 2019.

[54] LEMOS, S.G. C.; CAMPOS, D. G.; FERREIRA, C.S.C.; BALARO, M. F.A.; CUNHA, L. E.R.; PINNA, A. E.; BRANDÃO, F. Z. Vascularização uterina de éguas prenhes de jumento e de garanhão monitoradas por ultrassonografia Doppler. Pesquisa Veterinária Brasileira, v. 37, n. 8, p. 877-882, 2017.

[55] OLIVERIA, M.; OLIVEIRA, L. F. P. S.; LUBA, C.; CAMPOS, I. S.; FERREIRA, C. S. C.; SOUZA, G. N.; PINNA, A. E.; FERREIRA, A. M. R. Índices de resistência e pulsatilidade das artérias umbilicais em éguas gestantes de jumento vs garanhão nos primeiros 50 dias de gestação. Anais da XX Conferência Anual ABRAVEQ, São Paulo – SP, 2019.

[56] LEMES, K.M.; SILVA, L.A.; ALONSO, M.A.; CELEGHINI, E.C.C.; PUGLIESI, G.; CARVALHO, H.F.; AFFONSO, F.J.; SILVA, D.F.; LEITE, T.G.; ARRUDA, R.P. Uterine vascular perfusion and involution during the postpartum period in mares, Journal of Equine Veterinary Science, V. 51, p.61-69, 2017.

Section 3

Reproductive Techniques in Aquatic Animals

Chapter 5

Cryopreservation and Its Application in Aquaculture

Judith Betsy C, Siva C and Stephen Sampath Kumar J

Abstract

Aquaculture is the major aquatic animal production segment. Problems like inbreeding depression, genetic drift, introgressive hybridization, etc. have been influencing the production of quality seeds negatively. Cryopreservation serves as a way-out for these problems and a possible answer to produce quality seeds and genetically improved varieties. It has been considered as a major strategy for conservation of fish genetic resources. Cryopreservation of fish gametes has been in vogue since 1953 and the technology is well studied and validated for many species. So far the milt of 200 fish species has been cryopreserved successfully. In this chapter, the importance of aquaculture in overcoming malnutrition, genetic issues affecting quality seed production, cryopreservation protocol employed for various fish species, problems faced in cryopreserving fish eggs and embryos and future of cryopreservation in aquaculture have been discussed.

Keywords: Cryopreservation, aquaculture, fish, spermatozoa, egg, inbreeding

1. Introduction

In the World Summit on Food Security held at Rome during 16–18 November 2009, it was committed to eradicate hunger from earth by increasing investment in agriculture involving public and private enterprises. Food and Agriculture Organization (FAO) defined food insecurity as a situation that exists when people lack secure access to sufficient amounts of safe and nutritious food for normal growth and development and an active and healthy life.

The projections of FAO for the next 15 years indicate that, if agricultural innovation continues at a reasonable rate, food production can increase by 2 percent per year in the developing countries. Another report by World Bank mentioned that the world needs to produce at least 50% more food to feed 9 billion people by 2050. If the natural resources offer good potential for agricultural development, supporting agriculture research can bring big benefits in reducing food insecurity and malnutrition [1]. In solving the above issues, along with grains and vegetables, fish can also play a major role.

Fish and other aquatic products provide at least 20% of protein intake for one third of the world's population and the dependence on fish is high in developing countries [1]. Small-scale fisheries are considered to be more important for food security because they supply more than half of the protein and minerals for over 400 million people in the food deficit countries of Africa and South Asia [2].

Fish is not an energy food, but it is an essential food for the human being. It is an extremely important source of protein, minerals and oils in many under developed countries. Fish protein constitutes around 30% of the Micronesian diet and 15% of the Polynesian diet [2]. Fish is more nutritious than other staple foods, providing quality animal protein, essential fatty acids and micronutrients. Interventions related to fish intake and aquaculture production include, utilizing fish as complementary food to improve nutritional status of children, encouraging children and women to eat nutrient-dense fish through nutrition education at community level [3] and increasing production of more demanded fish species through effective dissemination of the technology [4].

Fish being an important food for mankind, its production through all possible means has been explored. Besides exploitation of natural resources like sea and natural water impoundments, culture production through structured methods deserve due attention now. This shift in the population in the natural waters has impacted the availability of fish through capture and driving the people to develop ways to produce the fish through aquaculture. Nevertheless, the impacts of climate change on aquaculture also could not be pushed off.

Studies in Asia suggest that low-income households consume lesser quantities of fish than rich households [5], but they still depend on fish as a major source of animal protein [6]. This invariably suggests that fish supply should be sufficient to wade off the malnutrition from the low-income group of people in order to keep the life free from nutritional disorders.

It is at this point anthropologists in the world insist that apart from concentrating on improving agriculture production, agencies like FAO and World Bank must establish projects in aquaculture of species that are positively impacted by climate change and at the same time contains high nutrients to eradicate malnutrition. Some of the fishes rich in nutrition include carps, catfish, murrels, tilapia and prawn [7].

Globally, fish provides 20% of average per capita intake of animal proteins to more than 3.3 billion people. In some countries like Bangladesh, Cambodia, the Gambia, Ghana and Indonesia, fish contribute to 50% or more of animal protein. The global fish consumption per capita food grew from 9.0 kg (live weight equivalent) in 1961 to 20.5 kg in 2018 which is about 1.5% increase per year [8].

The total world fisheries and aquaculture production has reached 179 million tonnes in 2018 which was recorded as highest of all times and estimated at USD 401 billion [8]. The aquaculture sector was the main driver that led to the increase in production of aquatic animals, and the total aquaculture production was estimated to be 82.1 million tonnes valued at USD 250 billion with average growth of 5.3% per year. The contribution of world aquaculture to global fish production has increased from 25.7% in 2000 to 46% in 2018.

2. Genetic issues affecting seed quality

Aquaculture in many countries depends on the adequate supply of quality seeds. However, it is not always possible since many hatcheries have reported stock deterioration due to poor brood stock management, inbreeding depression, genetic drift, introgressive hybridization, unconscious selection, lack of effective population size (N_e) and genetic erosion of domesticated stock.

2.1 Prevention of detrimental effect of inbreeding and genetic drift via cryopreservation

Inbreeding and genetic drift cause undesired changes in the genome and result in lower viability and growth and increase developmental instability in fishes [9].

Unplanned and uncontrolled breeding often lead to inbreeding depression which lead to decreased growth rate, low fecundity and poor survival which are usually accompanied by loss of alleles via genetic drift [10]. Uncontrolled inbreeding and genetic drift occur together in closed hatchery populations and these factors are determined by the population's N_e. Hence maintaining the desired N_e will prevent adverse effects on productivity and profits [10].

Due to high fecundity in fishes, inbreeding is more prevalent in aquaculture than other domesticated animals. This applies especially to highly fecund species like Indian Major Carps (IMC) (catla, *Catla catla,* rohu, *Labeo rohita,* mrigal, *Cirrhinus mrigala*) and Chinese carps (silver carp, *Hypophthalmicthys molitrix*, grass carp, *Ctenopharyngodon idella*, common carp, *Cyprinus carpio*) where few broodstock are necessary to meet demands for fry and broodstock replacement. The detrimental effects of inbreeding are well documented and can result in 30% or more decrease in growth, survival and reproduction [11].

The problem of inbreeding and genetic drift can be reduced by spawning more fish than needed. Since the fecundity is high in some species, the required number of fingerlings can be produced by breeding one or two females and males. But the ability to spawn relatively few fish must be moderated if inbreeding and genetic drift are to be controlled. Another way to increase N_e and reduce the rate of inbreeding and genetic drift is to spawn a more equal sex ratio. Most farmers and hatchery managers use skewed sex ratios when they spawn their fish. This is done because one male can usually be used to fertilize eggs from several females. This enables farmers to use and maintain fewer males, which lowers production costs. When a skewed sex ratio is used, the rarer sex has a disproportionate influence on the size of N_e. Cryopreservation can help in maintaining N_e by breeding more number of fish which leads to maintenance of fewer males in the hatchery [10].

2.2 Introgressive hybridization with cryopreservation

It is the gradual infiltration of the germplasm of one species into that of another as a consequence of hybridization and repeated backcrossing [12]. Sarder et al. [13] reported that pure mrigal are severely being threatened by introgressive hybridization in Bangladesh. Moreover, unplanned hybridization, inbreeding depression and genetic drifts have been the causes of deteriorating quality of this species. They opined that cryopreservation is the simplest and most inexpensive method to preserve genomes that can be used to maintain future conservation options. Introgression of autochthonous populations with introduced ones is a common phenomenon in salmonids and it can result in outbreeding depression and replacement of possibly locally adapted populations by allochthonous ones [14].

Horvath et al. [15] applied cryopreservation as a conservation effort of two salmonid species such as the marble trout (*Salmo marmoratus*) and the Adriatic lineage of the grayling (*Thymallus thymallus*) autochthonous to the drainage of the Soča river in Slovenia. Populations of these species were greatly affected by hybridization and introgression with allochthonous species such as the brown trout (*Salmo trutta m. fario*) and the Danubian lineage of the grayling that were introduced to the Soča drainage during the 20th century. Cryopreservation of sperm from the Adriatic grayling and the marble trout has constituted an integral part of the conservation activities. In case of the grayling, no pure population was available and hence the proportion of Adriatic genotype in the broodstock was increased. Genetic analyses of the populations were also conducted. Sperm and fin clips were collected from wild males on the spawning grounds. Sperm was cryopreserved and stored until the genetic analysis was completed on each sample. Cryopreserved sperm of individuals containing higher than a pre-defined proportion of Adriatic genotype was thawed and used for fertilization of eggs from Adriatic females. The resulting progeny was developed as broodstock and 70–80% of the local grayling broodstock originated from cryopreserved sperm. In case of the marble trout, cryopreservation was used to create "sanctuary" streams. Sperm is collected from wild males of a given pure population prior to the spawning season (early November) and cryopreserved. Sperm is stored in liquid nitrogen (LN_2) until the spawning season (December–January) and then eggs of females from the identical population are fertilized with the cryopreserved sperm. Eyed eggs are then stocked into artificially created nests in the prepared "sanctuary" stream. Thus, a high number of males of the given pure population participate in the creation of the new population.

From a management perspective, the desirability of introgressive hybridization in response to environmental change depends on the circumstances. It is desirable when the resulting adaptation has the potential to rescue a native species from extinction, such as adaptation to the sudden climate shifts that might become more frequent or extreme with climate change. In such cases, management actions to protect hybridization, such as the protection of hybrid zones, might enhance the potential for species to respond to environmental change [16].

3. Aquaculture and cryopreservation

Fish breeding depends on many factors and failure or partial success in the breeding is a reality for many successful hatchery operators. In order to get the required quantity of seeds, induced breeding is considered as a viable tool that makes the fishes maturing and spawning despite to low or poor rainfall and worst climatic conditions. Nevertheless, the health of brooders is severely affected by repeated breeding attempts within its confined life time. Exchange of brooders is not a simple task due to difficulties and physiological factors associated with the transportation of the brooders. Therefore shipping of gametes is considered as a possible alternative that may have its own advantages as witnessed in the animal husbandry.

It is necessary to introduce biotechnological tools in fish breeding programme to ensure continuous seed production. Cryopreservation may be a possible answer to produce quality seeds and genetically improved varieties. FAO has endorsed cryopreservation as a major strategy for conservation of fish resources [17]. Cryopreservation increases the longevity of gametes for several years without any drastic change in the fertilizing capacity of the gametes by lowering the temperature usually to −196°C [18] which arrests all biological activities, including biochemical reactions that lead to cell death and DNA degradation [19].

In fishes, Blaxter [20] is believed to be the first successful scientist who did the cryopreservation of herring spermatozoa and proceeded up to artificial fertilization with the cryopreserved spermatozoa. It has been reported that so far milt from over 200 species of freshwater and marine fish have been cryopreserved [21, 22].

4. Principle of cryopreservation

The basic principle of cryopreservation is exposure of living cells to sub-zero temperature as low as −200° C through a perfect process thereby arresting its activities without damaging the life of it. A series of complex and dynamic processes of heat and water transport between cells and their surrounding medium is involved during the freeze–thaw process of biological material. The effect of the process depends on the speed at which the cells are frozen or thawed. When cells are frozen in an aqueous solution, both cells and the solution get super cooled leading to freezing that will be followed by heterogeneous nucleation, usually in the extracellular solution. The same condition can be seen in the cell solutions also. If such condition occurs intracellularly, the resultant nuclei will be isolated by plasma membranes from the unfrozen cell components and leads to separation of ice crystals inside the cell. As water gets frozen, the extracellular solution becomes progressively more concentrated leading to slow dewatering conditions in the cells. This results only when the cooling is slow and there is sufficient time for the cells to lose enough water so as to remain in osmotic equilibrium with the concentrating extracellular solution leading to water loss inside the cells. If that occurs, that will lead to cell death otherwise called as freeze killing or chill killing. While this may take time in large and multi cellular organisms, in small micro-organisms and single cells much of water can be withdrawn during freezing leading to desiccation and the death of the cell instantly [23].

In contrary to the above situation, if the rate of cooling is faster and rapid, there will be less time for the intracellular water to diffuse out of the cells. A balancing situation will emerge under such fast or rapid cooling. This leads to survival of the cells by minimizing the time or exposure duration to concentrated solution. The cooling rate also ensures there is no formation of intracellular ice. This process is called vitrification and it is the process that is happening inside the cell in the cryopreservation process [23].

During thawing, the same cellular physiological processes occur in reverse order. The thawing rate should also be rapid and fast enough that of the corresponding cooling rate. Nevertheless, recrystallization invariably occurs during thawing, forming lethal intracellular ice. A high warming rate is usually employed to minimize the degree of recrystallization when thawing is rapid to provide insufficient time or least possibility for the dehydrated cells to absorb the amount of water lost during freezing [23].

5. Cryopreservation of fish spermatozoa

5.1 Milt collection

Cryopreservation success depends on the milt quality and hence, quality of milt must be evaluated based on the condition of spermatozoa prior to cryopreservation. Milt should be always collected from oozing ripe brooders by stripping method in ice cold, sterilized cryovials [19]. Milt must be collected in clean, dry and sterile vials and immediately stored on ice [19]. Collected milt should be in quiescent form and should be free from contaminants, such as water, mucus, blood, and gut exudates. Prior to stripping, the urinary bladder can be emptied by gentle squeezing in order to avoid milt contamination with urine.

Employing stripping method for collecting fish milt might result in contamination with urine which may seriously influence milt characteristics and quality [24]. The contaminated milt can deteriorate the spermatozoa quality and have detrimental effects on post thaw viability as the contaminants such as urine, blood, mucus, etc. can change the seminal fluid composition and induce sperm motility [25]. Urine contamination can lead to lower percentage of fertilized eggs [26].

Using a catheter for milt collection can avoid urine and fecal contamination [27–29]. Researchers have suggested that anesthetizing the donors prior to milt collection was advantageous [30, 31]. Anesthetizing agents like Tricaine methane sulphonate (MS-222) can be used before milt collection [32]. Fish can be anesthetized by immersing in 2-phenoxyethanol for 2 min at a dose of 0.5 ml/l of water [33].

For instance, *O. mykiss* was anesthetized with MS 222 in a 1:10,000 dilution water bath during milt collection and the milt was collected by gently massaging the abdomens of the fish [34]. *C. carpio* brooders were anesthetized with a 1:1000 aqueous solution of 2-phenoxyethanol before handling [35]. *C. carpio* males were anesthetized with 2-phenoxyethanol at a dose of 0.5 ml before milt collection [36].

5.2 Spermatological properties

Sperm quality evaluation is very important as it provides necessary information for optimal handling and storage protocols for sperm used in artificial fertilization [25, 37]. The fish milt composition and its physical characteristics vary with species and are important from the aspect of milt quality [38]. The quality of the milt is species specific [39] and can be affected by the feeding regime, feed quality, rearing temperature and spawning season of males [40, 41]. Spermatozoa motility, milt volume and the spermatozoa concentration are considered to be good indicators for milt quality [42, 43]. Sperm quality can be evaluated based on the sperm volume, spermatozoa density, motility of spermatozoa [25]. The appearance, color and nature of milt are also used to assess the quality of milt. The milt volume of fishes is found to vary with species [44–46].

5.2.1 Sperm motility and motility duration

Motility is one of the most important parameters which is most frequently used to assess milt quality after cryopreservation and generally presents a positive correlation with fertilizing capacity [47]. Sperm motility is considered as the best biomarker of milt quality [48]. Motility depends on various aspects of the cell, such as the physiological state of the mitochondria, ATP production, plasma membrane channel integrity and flagellum structure [49]. Relationship between percentage motility and fertilization capacity of spermatozoa was reported in many fishes [50–55]. The motility, velocity and fertilizing ability of sperm was found to vary according to seasonal variations in osmolality of seminal plasma [56–58]. Sperm motility was also found to vary in vigor and duration among individual male depending on ripeness [45].

The spermatozoa are in immobile phase before ejaculation and it was reported that the osmolarity and ion content of the aquatic medium are central factors in activating motility [59, 60]. It was observed that in some of the fish species, the changes in the osmotic pressure (0–300 mosmol/l) could initiate spermatozoa motility [61]. In carp testes and seminal plasma, inhibition of sperm motility was observed due to high osmolality (approximately 300 mosmol/kg) surrounding spermatozoa [50]. Various researchers observed that the spermatozoa usually remain motile for less than 2 min and sometimes they are only highly active for less than 30 s in most of the freshwater fishes [62–64].

5.2.2 Sperm pH

Milt pH can affect spermatozoa motility and maturation [25]. Hence determination of variation in sperm pH provides information on fertilization capacity of spermatozoa. The milt of most of the freshwater fish species exhibit slightly alkaline pH [65]. When intracellular pH is below 7.5, sperm cells remain immotile with low respiration rate, but in response to an internal alkalinisation, they become motile, concomitantly with an increase in oxygen consumption [66, 67]. The initiation and duration of sperm motility is influenced by the extracellular and intracellular pH [68]. The external pH affects intracellular proton concentration which modifies the membrane potential and motility behavior [69].

5.2.3 Sperm density

Traditionally the density of sperm has been used for the assessment of milt quality. It is an important parameter which has an impact on fertilization success and is a characteristic feature of fish species [70]. Spermatozoa density is usually reflected by sperm volume [71]. Various methods like using Sysmex Microcell counter CC-120 [72], spectrophotometric method [73], haemocytometric method [74] were employed to estimate sperm density in fishes.

5.3 Extender

For successful cryopreservation, it is essential to prevent activation of spermatozoa during preservation. Undiluted milt is unsuitable for storage at cryogenic temperatures, so it should be diluted with an appropriate medium [51]. Because motility of fish spermatozoa is mostly a one-time event, this medium should not induce motility and at the same time must not interfere with the ability of the spermatozoa to be activated subsequently during utilization. Media that satisfies these conditions is called "Extender" [75].

Extender is a salt solution which helps to maintain the viability of cell during cryopreservation, which supplies sources of energy to sperm cells, protect the cells from temperature related damage, and maintain a suitable environment for the sperm to survive during the period of cryopreservation [76, 77]. Based on the inorganic composition of seminal plasma, extender is prepared as a buffered physiological saline solution [78] and hence extender composition differs between species. Extenders maintains the inactivity of spermatozoa when milt is diluted before freezing due to stabilization of physicochemical properties [79]. Sperm typically need to be maintained in an extender with proper osmolality (usually nearly isotonic to the plasma osmolality) to inhibit undesired sperm activation during refrigerated storage or cryopreservation [80].

A large number of extenders such as Ringer's solution, Cortland's solution, Alsever's solution, etc. have been tried for the cryopreservation of spermatozoa of fish which were proven successful for milt cryopreservation in mammals [29]. Several simple extenders which are isotonic in nature, with inorganic salts like NaCl, KCl, $CaCl_2$, $NaHCO_3$, $NaHPO_4$, $MgSO_4$, $MgCl_2$ and others with organic compounds such as fructose, mannitol, lecithin, glycine have been used with varying levels of success [29]. Extenders have been developed using saline and sugar-based diluents [81].

Tris-egg yolk gave higher post-thaw motility percentage (50%) during cryopreservation of milt of *C. carpio* and *L. rohita* [82]. Use of glucose-based extender containing 10% dimethyl sulfoxide (DMSO) could be successfully used for *Oncorhynchus mykiss* milt cryopreservation and fertilization rate similar to that of fresh spermatozoa can be achieved [45]. Sperm diluted with 0.3427 g NaCl, 3.4314 g sucrose, 100 ml DW, 21 µl NaOH solution, 0.5 ml antibiotic (10,000 unit/ml penicillin and 10,000 µg/ml streptomycin) and DMSO gave the best post-thaw motility (94.5 ± 3.3%) in *C. carpio* [83]. The feasibility of three extenders namely, Freshwater Fish Saline, Modified Fish Ringer and Physiological Saline was compared in cryopreserving *C. carpio* milt and the motility duration obtained was 57.28 ± 9.21 s, 64.78 ± 8.84 s and 67.39 ± 4.79 s for Physiological saline, Freshwater Fish saline and Modified Fish Ringer respectively [84].

5.4 Cryoprotectant

Cryoprotectants are low molecular weight compounds that penetrate cells and lower the freezing points of solutions. Cryoprotectants in combination with an effective dilution ratio can also improve the cryo-resistance of spermatozoa [85]. Cryoprotectants need time to penetrate to the cells (equilibration), however, prolonged exposure before cryopreservation can be toxic for sperm [85]. At higher concentrations, cryoprotectants can suppress most of cryoinjuries but at the same time, it can become toxic to the cells [86]. Therefore, suitable cryoprotectant concentration is needed for the development of cryopreservation protocol. The protective effect of cryoprotectants varies in different fish species [87].

Cryoprotectants are very essential for the survival of spermatozoa during cryopreservation. There are two different types of cryoprotectants; permeating and non–permeating [88]. Permeating cryoprotectants such as DMSO, glycerol, methanol, propanediol etc., are believed to lower the freezing point of the solution, which minimize osmotic shock by replacing the water inside the cell, and reduce formation of destructive intracellular ice [89]. Non-permeating cryoprotectants include protein like milk, egg yolk, bovine serum albumin (BSA); sugars such as glucose, sucrose; synthetic polymers like polyethylene glycol and polyvinylpyrrolidone and are believed to stabilize the membrane during cryopreservation [90]. Use of insufficient cryoprotectant before cooling reduces effectiveness, whereas excessive cryoprotectant causes osmotic swelling and rupture during thawing and dilution [91]. Cryoprotectants were found to prevent the formation of ice crystals during freezing [48].

Due to ice crystal formations at low temperatures very few spermatozoa survive without cryoprotectant and same levels of those cryoprotectants can be lethal to unfrozen cell [92]. Cryoprotectants were most effective when they could rapidly penetrate the cell during freezing, which resulted in delay in intracellular freezing and led to minimization of the solution effect [93]. Common cryoprotectants used for fish sperm include DMSO, methanol and propylene glycol (PG) [94].

Regarding these cryoprotectants, PG used for sperm cryopreservation in yellowtail flounder (*Pleuronectes ferrugineus)* resulted to be an effective cryoprotectant [95] but showed moderately good post-thaw motility in *Clarias gariepinus* [96]. Methanol at 10% was found suitable for cryopreservation of bitterling milt [97], bagrid catfish [98] and *C. gariepinus* [99] and 5% methanol was reported to be suitable for tilapia (*Oreochromis niloticus*) milt cryopreservation [94]. DMSO was established to be very successful for cryopreservation of sperm in various freshwater species [100, 101] and has been considered as a universal cryoprotectant [102, 103].

The milt of *C. mrigala* when cryopreserved with glucose as co-cryoprotectant at 0.5% concentration egg yolk at 10% concentration gave the highest post-thaw motility duration [104, 105]. BSA at 2% gave the highest post-thaw motility duration in *C. carpio* [106].

5.5 Dilution ratio

The process of milt dilution is carried out as a means to increase the number of eggs that can be fertilized with a small volume of milt [107]. In fish spermatozoa cryopreservation, dilution of the sperm fluid is one of the most important steps which has been reported to improve fertilization rate as compared with results obtained with undiluted milt [108]. Milt dilution ratio is very important for fish sperm to survive after cryopreservation [109]. The dilution process is very important to increase the volume of milt, so that it can be used for multiple inseminations. Milt is generally diluted 3–20 folds for Salmonid, carp and tilapia [24].

In Cyprinids, full sperm motility is activated at osmolalities <50 mosmol/kg [50]. Using cryopreserved milt, full activation of sperm motility was obtained at ratios of milt to fertilization media of 1:10 for all types of media since at this ratio, the osmolality of the extender–water mixture was high enough to stabilize sperm viability [110]. Too low dilution ratio do not activate full sperm motility and too high ratios results in insufficient low sperm concentrations in the fertilization solution [110]. However, reports also suggest 1:25 [111, 112] and 1:20 [113] as the optimal ratio of milt to fertilization medium.

In this regard, when *C. carpio* milt was diluted with Kurokura medium at 1:5 ratio, it gave best results [114]. Dilution ratio of milt to extender of 1:7 resulted in highest hatching rates while at lower (1,3) and higher dilutions (1,10) fertility was significantly decreased in bleak (*Chalcalburnus chalcalburnus*) [110]. When dilutions of 1:25, 1:50 and 1:100 were evaluated on European perch (*Perca fluviatilis*), best result was obtained at 1:50 dilution [115]. The highest mean post-thaw motility duration, motility score, percentage of fertilized eggs, and hatching rate was obtained with 1:40 dilution ratio in *C. carpio* [116].

5.6 Equilibration period

Equilibration period is the optimum time that must be allowed to facilitate the penetration of permeating cryoprotectants into the cells while minimizing the toxicity for effective protection during freezing [117].During cryopreservation of milt, an equilibration time of 45–60 min for IMC [118], 10 min for *C. carpio* [119], 60 min for *Tor putitora* [120], 5 min for bleak (*C. chalcoides*) [100], 10 min for *L. rohita* [121], 10 min for *O. mykiss* [122] has been employed with successful results.

Salmo gairdneri milt stored for 20 min after dilution gave significantly higher percentage of fertilization than that stored for 65 min or longer and therefore, cryopreservation of milt should be done as soon as possible after the collection [123]. Poor post-thaw motility was recorded in *C. carpio* at 20 min of equilibration time than that frozen immediately [124]. When the diluted milt was equilibrated for 15 min, there was no adverse effect on the post-thaw fertility of Salmonid milt [48].

5.7 Freezing

Too high freezing rate result in the formation of small ice crystals within the cell due to limited time for the free water to separate from the cytoplasm which puncturescell membraneand the membranes of the cell organelles. Too low freezing rate exposes the cell to the concentrated cytoplasm for a long time resulting in pickling effect and the biomolecules in the cell get denatured due to the high salt concentration and subsequent changes in the pH [18].

The optimum freezing rate is a moderate rate between the two extremes of the freezing rate [125] which depends on cell type and size, cryoprotectant type and concentration, equilibration time, final temperature prior to plunging in LN_2, fish species and associated interactions [99, 126]. Optimal cooling rate should be rapid enough to minimize the duration of exposure to prevent the occurrence of concentrated solute and slow enough to allow water osmosis to prevent intracellular ice crystal formation [127].

The freezing rate is a critical factor and it was reported that instant immersion in LN_2 may significantly decrease the post-thaw motility duration of fish spermatozoa [128]. Freezing can be performed by programmable temperature changes or simple immersion in LN_2 vapor above the surface of LN_2 [83]. Freezing can also be done using methanol-dry ice bath [129] or by freezing the extended milt with cryoprotectant over crushed dry ice [130]. The pelletization technique in which specific volumes of diluted milt is placed over dry ice (solid CO_2) also served to freeze the milt and it was used by many workers [32, 131, 132].

In a protocol, straws were frozen for 4 min on a stainless steel tray (−80°C) suspended over LN_2 and was immersed into LN_2 [133]. When *C. carpio* milt was frozen 3 cm above the surface of LN_2 for 3 min before plunging in LN_2 it resulted in high post-thaw motility as well as fertilization and hatching rate [134]. During cryopreservation of *C. carpio* milt, the 0.5 ml straws were placed horizontally onto a 3 cm high styroframe raft (−130°C) for 20 min, which was floating on the surface of LN_2, before immersing the straws into LN_2 and it did not negatively affect the fertility of frozen–thawed sperm [35].

Programmable freezers was also used by many researchers for freezing the diluted milt samples of several fish species [135–137]. Programmable freezing allows the pre-setting of different freezing programs, the monitoring of precise temperature during the cooling sections and the continuous biological examination of cells during the freezing stages [138]. Different programmes and different final temperatures can be attained in programmable freezer [119, 139–141]. The use of programmable freezer allows the evaluation of spermatozoa motility at different rates of cooling during freezing [141]. Incorporation of fast freezing rates using the controlled-rate programmable freezer was successfully used in earlier studies for cryopreservation of carp sperm [111, 113, 142].

For cryopreservation of milt of *C. carpio,* the most efficient freezing rate was 5°C/min from 2°C to −7°C and 25° C/ min from −7° C to −70°C [135]. A slower cooling rate at 4°C/min from 0^0 C to −4°C and 11°C/min from −4°C to −80°C can also be used for cryopreservation of *C. carpio* milt successfully [113]. A cooling program of 4°C to −9°C at a rate of 4°C/min and then from −9° C to −80° C at a rate of 11°C/min, which was held for 6 min at −80°C, and transferred into LN_2 was followed for *C. carpio* and high motility (69 ± 14%) and moderate fertilization rate (56 ± 10%) was reported [111]. *C. gariepinus* spermatozoa can be frozen at the rate of −5°C/min initially from +5°C to −35°C and then from −35°C to −50°C or − 70°C [99].

Three different cooling methods were employed during the cryopreservation of *C. carpio* [142]. Two of them used 3 steps, initially from 2°C to −7°C then −7°C to −30°C and finally −30°C to −80°C with two different cooling rates (3 and 6°C/min) after which the sample was transferred to LN_2. In the third method, a one-step method (2°C to −50°C) with faster average cooling rate (10°C/min) was applied and was reported that faster cooling rates (6 and 10°C/min) were more efficient for cryopreservation and the highest fertilization recorded with 10°C/ min was 99%.

5.8 Thawing

The rate of thawing is an important step which is said to be a decisive factor for the success of cryopreservation procedure. It is the reverse of freezing but rapid thawing after the cooling procedure is preferred however, too high and too low rates of thawing are detrimental for the cryopreserved spermatozoa [18]. Thawing rates should be high enough to avoid recrystallization as its rate is very critical for preservation of spermatozoa viability [143]. It appeared that the ideal thawing procedure almost avoided or reduced either recrystallization and ice crystal formation during thawing. The temperature change should allow movement of water and cryoprotectants while preventing intracellular ice recrystallization [144].

In Cyprinid fishes, the highest mean fertilization percentage of 57% was obtained in *C. idella* when thawed at 20°C quickly in a water bath [145]. The cryopreserved milt of freshwater carps (*L. rohita*, *C. carpio*, *Puntius gonionotus, C. idella*, *Aristichthys nobilis* and *Pangasius sutchi*) was thawed by swirling the frozen ampoules in tap water at 29°C [146]. Similarly, the frozen milt of IMC and *H. molitrix* was thawed by swirling the straws in tap water at 30°C [147]. High post-thaw motility percentage of 92–98% and high hatching percentage of 25.7% was obtained after thawing the cryopreserved milt of *T. khudree* at 37 ± 1°C for 5–10 s [148]. The highest mean motility (83.4 ± 2.1) and fertilization rate (85.6 ± 2.8) was obtained in *C. Idella* when the milt was thawed at 35° C for 30 s [149]. The highest post-thaw motility of 52.6 ± 1.4 s was recorded in *C. carpio* when thawed at 30°C for 30 s [150].

In Salmonid fishes (*O. mykiss, Salmo trutta lacustris, S. trutta fario* and *Salvelinus fontinalis*), the highest fertilization rates obtained was when milt was thawed at 25°C in water bath for 30 s and change of the thawing period for only 5 s or the thawing temperatures for 5°C led to reduce of post-thaw fertilization ability of milt [45]. Cryopreserved milt of *T. khudree* was thawed at 37°C for 40 s in a water bath [120]. Cryopreserved milt of *O. mykiss* was thawed at 25°C in water bath for 30 s for 0.5 ml and 1.8 ml straws and at 60°C for 30s/ 80°C for 20 s for 5 ml straws and was reported that thawing at 25°C in water bath for 30 s was best for thawing of *O. mykiss* milt [42]. The cryopreserved milt of *O. mykiss* was thawed at 10°C for 30s in water bath [130]. The cryopreserved milt of Salmonid fishes was thawed at 25°C for 30 s for 0.5 ml straws and at 30°C for 30 s for 1.2 ml and 5.0 ml straws in a water bath [151].

6. Cryopreservation of fish eggs and embryos

Cryopreservation of fish eggs and embryos are still in its infant stage. Unlike cryopreservation of spermatozoa, very few studies are available on cryopreservation of eggs. Attempts have been made to cryopreserve the eggs of rainbow trout [152, 153] and embryos of Japanese medaka fish, *Oryzias latipes* [154], rainbow trout, *O. mykiss* [155], zebra fish, *Brachydanio rerio* [156], common carp, *C. carpio* [157] and rohu, *L. rohita* [158].

Many of the attempts to cryopreserve fish eggs were failure due to dehydration, relatively large size of eggs, presence of large amount of yolk and different water permeability rate of membranes [159–161]. The major hindrances recorded in the cryopreservation of egg and embryos of teleost fishes [162] are the large size of fish egg and embryos which results in low surface/volume ratio and lower membrane permeability to water and cryoprotectant solutions that makes the embryos difficult to cool and warm uniformly without damage and ice formation, low permeability of the membrane due to the presence of chorionic layer, sensitivity of fish egg and embryosto low temperatures and the presence of multi-layered membrane structure which hinders the osmotic properties for each compartment of the egg/embryos which finally affects the transport of the cryoprotectant solutions.

Studies have been carried out by different researchers to overcome these issues and some of the efforts made are microinjection of cryoprotectants directly into the cytoplasm [163], use of negative pressure on the egg/embryos to increase permeability of the cryoprotectants [164], microinjection of anti-freeze protein [165] and application of hydrostatic pressure on the egg/embryos [166]. Precise knowledge of embryo permeability is essential for successful cryopreservation of egg/embryos [167].

Herring embryos did not survive after cooling below −10°C when DMSO was used [168]. Methanol was a better cryoprotectant for zebrafish embryo when compared with DMSO or ethanediol since it penetrates theentire embryo within 15 min while other cryoprotectants could not penetrate into yolk even after 2.5 h [169, 170]. Similarly, PG also could not protect the zebrafish embryos upon immersion of it into LN_2 as it resulted in mitochondrial damage, disorganization of ribosomes and plasma membrane of the yolk syncytial layer [171].

7. Application of cryopreservation in aquaculture

- This technology can be used to preserve milt of the best age group brooder which can be used at any point of time in future.
- It can also eliminate inbreeding problem since cryopreserved spermatozoa can be easily exchanged between hatcheries.
- Using this technology, spermatozoa can be made available at any season of the year.
- It makes breeding possible during off-season.
- It synchronizes the gamete availability of both sexes leading to sperm economy.
- It simplifies broodstock management in farms.
- It helps in the production of viable and strong offspring by intra-species hybridization.
- It overcomes the difficulties arising due to the short time viability of gametes.
- It enables the genetic preservation of desired lines.
- It allows cross breeding at different times of the year.
- It helps in germplasm storage for genetic selection programs or conservation of species.
- Cryopreserved spermatozoa can help in the hybridization programmes and genetic engineering research in fishes.
- It leads to many other avenues such as cryobanking of viable gametes as in the case of animal production and development of gene bank and genetic manipulation in fishes.

8. Demerits of frozen milt in aquaculture

- All the milt collected from individuals do not withstand rigors of freezing
- High initial investment cost
- Limits number of sires/males used and if proper care is not taken it may lead to inbreeding
- Requires better training of personnel
- Reduced or poor fertilization rate compared to other artificial breeding methods

9. Conclusion

Cryopreservation technology has been developed for many fish species. However, standard species specific cryopreservation protocols must be developed and the success rate of using cryopreserved sperm in artificial fertilization program of every fish species has to be determined for commercializing the technology. Even though standard protocols of cryopreservation are followed, cryoinjuries are unavoidable. Ways to overcome the cryoinjuries by establishing proper freeze–thaw cycle is essential. The oxidative stress in the cryopreserved sperm must be clearly addressed and methods to reduce the production of reactive oxygen species (ROS) must be evolved. The possible effects of cryopreservation on the energy production, ROS production, mitochondrial DNA of the spermatozoa and the structure of spermatozoa must be documented. Unlike in animals, very few fish sperm banks have been established for fishes. More research is needed to make the sperm banks for fishes a reality in the developing countries. Addressing the research needs mentioned above will help to establish successful fish sperm banks for many commercially important fish species.

Conflict of interest

The authors declare that they have no conflict of interest.

Author details

Judith Betsy C[1]*, Siva C[2] and Stephen Sampath Kumar J[3]

1 TNJFU - Fisheries College and Research Institute, Tuticorin, Tamil Nadu, India

2 Molecular Genetics Laboratory, ICAR-Directorate of Coldwater Fisheries Research, Nainital, Uttarakhand, India

3 TNJFU - Directorate of Sustainable Aquaculture, Thanjavur, Tamil Nadu, India

*Address all correspondence to: betsy@tnfu.ac.in

References

[1] Béné C, Macfadyen G, Allison EH. Increasing the contribution of small-scale fisheries to poverty alleviation and food security. Food & Agriculture Org.; 2007.

[2] Beare D. Fisheries and aquaculture: The importance of fish for food and nutritional security. Beare, D., 2012. Fisheries and aquaculture: The importance of fish for food and nutritional security. In: Thornton, P. and Cramer, L. (Eds.), Impacts of climate change on the agricultural and aquatic systems and natural resources within the CGIAR's mandate. CCAFS Working Paper 23. CGIAR Research Program on Climate Change, Agriculture and Food Security (CCAFS). Copenhagen; 2012. pp. 55-67.

[3] Roos N, Wahab MA, Chamnan C, Thilsted SH. The role of fish in food-based strategies to combat vitamin A and mineral deficiencies in developing countries. The journal of Nutrition. 2007;137(4):1106-9.

[4] Thilsted SH, Roos N, Hassan N. The role of small indigenous fish species in food and nutrition security in Bangladesh. Naga, the ICLARM quarterly. 1997;20(3-4):82-4.

[5] Hossain M, Naher F, Shahabuddin Q. Food security and nutrition in Bangladesh: progress and determinants. eJADE: electronic Journal of Agricultural and Development Economics. 2005;2(853-2016-56126):103-32. DOI: 10.22004/ag.econ.110131

[6] Islam FU. Self-recruiting species (SRS) in aquaculture: their role in rural livelihoods in two areas of Bangladesh [thesis]. University of Stirling; 2007.

[7] Kawarazuka N. The contribution of fish intake, aquaculture, and small-scale fisheries to improving nutrition security: a literature review. 2010: p. 51

[8] FAO. The State of World Fisheries and Aquaculture. Food and Agriculture Organization of the United Nations, Rome. 2020: p. 206.

[9] FAO. Aquaculture development. 3. Genetic resource management. FAO Technical Guidelines for Responsible Fisheries. No. 5, Suppl. 3. Rome, FAO. 2008: p. 125.

[10] Tave D. Inbreeding and brood stock management. *Fisheries Technical Paper*. No. 392. Rome, Food & Agriculture Org.; 1999: p. 122.

[11] Dunham RA, Majumdar K, Hallerman E, Bartley D, Mair G, Hulata G, Liu Z, Pongthana N, Bakos J, Penman D, Gupta M. Review of the status of aquaculture genetics. In Aquaculture in the Third Millennium. Technical Proceedings of the Conference on Aquaculture in the Third Millennium, Bangkok, Thailand 2000: pp. 137-166.

[12] Anderson E. Introgressive hybridization. Biological Reviews. 1949:28(3),pp.280-307. doi:10.1111/j.1469-185x.1953.tb01379.x

[13] Sarder MR, Rafiquzzaman SM, Sultana R, Islam MF. Cryopreservation of spermatozoa of Mrigal, *Cirrhinus cirrhosus* with a view to minimize inbreeding and hybridization. Journal of the Bangladesh Agricultural University. 2009;7(1):211-218. DOI: 10.3329/jbau.v7i1.4986

[14] Horváth Á, Jesenšek D, Csorbai B, Bokor Z, Raboczki É, Kaczkó D, Bernáth G, Hoitsy G, Urbányi B, Bajec SS, Snoj A. Application of sperm cryopreservation to hatchery practice and species conservation: A case of the Adriatic grayling (*Thymallus thymallus*).

Aquaculture. 2012 Aug 15;358:213-215. doi:10.1016/j.aquaculture.2012.07.012

[15] Horvath A, Bokor Z, Bernath G, Urbanyi B, Snoj A, Susnik Bajec S, Jesensek D. Application Of Sperm Cryopreservation To The Culture And Conservation Of Salmonid Species: A Slovenian-Hungarian Collaboration. Serbia, 2015.

[16] Baskett ML, Gomulkiewicz R. Introgressive hybridization as a mechanism for species rescue. Theoretical Ecology. 2011;4(2):223-239. DOI 10.1007/s12080-011-0118-0

[17] Khoshoo TN. Conservation of India's endangered mega animals: Tiger and lion. Current Science. 1997;73(10): 830-42.

[18] Diwan AD, Ayyappan S, Lal KK, Lakra WS. Cryopreservation of fish gametes and embryos. The Indian Journal of Animal Sciences. 2010;80(4): 109-24.

[19] Agarwal NK. Cryopreservation of fish semen. Himalayan Aquatic Biodiversity Conservation & New Tools in Biotechnology. Transmedia Publication, Srinagar (Garhawal) Uttarakhand, India. 2011:104-27.

[20] Blaxter JH. Sperm storage and cross-fertilization of spring and autumn spawning herring. Nature. 1953;172 (4391):1189-90.

[21] Blesbois E, Labbé C. Main improvements in semen and embryo cryopreservation for fish and fowl. In Atelier sur la Cryoconservation des Ressources Génétiques Animales en Europe. Bureau des Ressources Génétiques. Paris. 2003. ISBN 2-908447-25-8: 55-65.

[22] Hiemstra SJ, van der Lende T, Woelders H. The potential of cryopreservation and reproductive technologies for animal genetic resources conservation strategies. InThe role of biotechnology in exploring and protecting agricultural genetic resources, FAO, 2006. pp.45-60.

[23] Kumar JSS and Betsy CJ. Cryopreservation of fish gametes and its role in enhancing aquaculture production. In: Santhanam P, Thirunavukkarasar AR, Perumal P, editors. Advances in Marine and Brackish water Aquaculture. India: Springer Publishers; 2015. p. 241-246.

[24] Rana KJ. Cryopreservation of fish spermatozoa. In Cryopreservation and Freeze-Drying Protocols. Humana Press, Totowa, NJ. 1995. pp. 151-165.

[25] Billard R, Cosson J, Perchec G, Linhart O. Biology of sperm and artificial reproduction in carp. Aquaculture. 1995;129(1-4):95-112.

[26] Bokor Z, Müller T, Bercsényi M, Horváth L, Urbányi B, Horváth Á. Cryopreservation of sperm of two European percid species, the pikeperch (*Sander lucioperca*) and the Volga pikeperch (*S. volgensis*). Acta Biologica Hungarica. 2007;58(2):199-207.

[27] Alderson R, Macneil AJ. Preliminary investigations of cryopreservation of milt of Atlantic salmon (*Salmo salar*) and its application to commercial farming. Aquaculture. 1984;43(1-3):351-4.

[28] Cabrita E, Alvarez R, Anel L, Rana KJ, Herraez MP. Sublethal damage during cryopreservation of rainbow trout sperm. Cryobiology. 1998;37(3): 245-53.

[29] Rao B, Soufir JC, Martin M, David G. Lipid peroxidation in human spermatozoa as relatd to midpiece abnormalities and motility. Gamete research. 1989;24(2):127-34.

[30] Kurokura H. Cryopreservation of rainbow trout sperm. Nippon Suisan

Gakkishi. *Bulletin* of the *Japanese Society* of *Scientific Fisheries.* 1980;46:1493-5.

[31] Lakra WS. Cryogenic preservation of fish spermatozoa and its application to aquaculture. Indian Journal of Cryogenics. 1993; 18, pp. 171-176.

[32] Piironen J. Cryopreservation of sperm from brown trout (*Salmo trutta* m. lacustris L.) and Arctic charr (*Salvelinus alpinus* L.). Aquaculture. 1993;116(2-3):275-85.

[33] Perchec G, Jeulin C, Cosson J, Andre F, Billard R. Relationship between sperm ATP content and motility of carp spermatozoa. Journal of Cell Science. 1995;108(2):747-53.

[34] Bozkurt Y, Akcay E, Tekin N, Secer S. Effect of freezing techniques, extenders and cryoprotectants on the fertilization rate of frozen rainbow trout (*Oncorhynchus mykiss*) sperm. *Israeli J. Aqua.– Bamidgeh*. 2005;57:125-130.

[35] Boryshpolets S, Dzyuba B, Rodina M, Li P, Hulak M, Gela D, Linhart O. Freeze-thawing as the factor of spontaneous activation of spermatozoa motility in common carp (*Cyprinus carpio* L.). Cryobiology. 2009;59(3):291-6.

[36] Cejko BI, Krejszeff S, Sarosiek B, Żarski D, Judycka S, Kowalski RK. Biochemical factors of common carp *Cyprinus carpio* L. 1758, seminal plasma and its relationship with sperm motility parameters. Journal of Applied Ichthyology. 2015;31:10-7.

[37] Linhart O, Gela D, Rodina M, Kocour M. Optimization of artificial propagation in European catfish, *Silurus glanis* L. Aquaculture. 2004;235(1-4):619-32.

[38] Kruger JD, Smit GL, Van Vuren JH, Ferreira JT. Some chemical and physical characteristics of the semen of *Cyprinus carpio* L. and *Oreochromis mossambicus* (Peters). Journal of Fish Biology. 1984;24(3):263-72.

[39] Piironen J. Variation in the properties of milt from the Finnish landlocked salmon (*Salmo salar* m. sebago Girard) during a spawning season. Aquaculture. 1985;48(3-4):337-50.

[40] Bromage N, Roberts RJ. Broodstock management and seed quality-general considerations. Brood-stock management and egg and larval quality. 1995:1-24.

[41] Rurangwa E, Kime DE, Ollevier F, Nash JP. The measurement of sperm motility and factors affecting sperm quality in cultured fish. Aquaculture. 2004;234(1-4):1-28.

[42] Cabrita E, Robles V, Alvarez R, Herráez MP. Cryopreservation of rainbow trout sperm in large volume straws: application to large scale fertilization. Aquaculture. 2001;201(3-4):301-14.

[43] Tekin N, Secer S, Akcay E, Bozkurt Y. Cryopreservation of rainbow trout (*Oncorhynchus mykiss*) semen. 2003; 55(3), pp. 208-212.

[44] Baynes SM, Scott AP. Seasonal variations in parameters of milt production and in plasma concentration of sex steroids of male rainbow trout (*Salmo gairdneri*). General and comparative endocrinology. 1985;57(1):150-60.

[45] Lahnsteiner F, Patzner RA, Weismann T. Semen cryopreservation of salmonid fishes: influence of handling parameters on the postthaw fertilization rate. Aquaculture Research. 1996;27(9):659-71.

[46] Lin F, Ciereszko A, Dabrowski K. Sperm production and cryopreservation in muskellunge after carp pituitary extract and human chorionic

gonadotropin injection. The Progressive Fish-Culturist. 1996;58(1):32-7.

[47] Kime DE, Van Look KJ, McAllister BG, Huyskens G, Rurangwa E, Ollevier F. Computer-assisted sperm analysis (CASA) as a tool for monitoring sperm quality in fish. Comparative Biochemistry and Physiology Part C Comparative Pharmacology and Toxicology, 2001;130, pp.425-433.

[48] Lahnsteiner F, Mansour N, Berger B. Seminal plasma proteins prolong the viability of rainbow trout (*Oncorynchus mykiss*) spermatozoa. Theriogenology. 2004;62(5):801-8.

[49] Cosson J, Groison AL, Suquet M, Fauvel C, Dreanno C, Billard R. Studying sperm motility in marine fish: an overview on the state of the art. Journal of Applied Ichthyology. 2008;24(4):460-86.

[50] Morisawa M, Suzuki K, Shimizu H, Morisawa S, Yasuda K. Effects of osmolality and potassium on motility of spermatozoa from freshwater cyprinid fishes. Journal of Experimental Biology. 1983;107(1):95-103.

[51] Stoss J. 6 Fish Gamete Preservation and Spermatozoan Physiology. Fish physiology. 1983;9:305-50.

[52] Harvey B, Kelley RN, Ashwood-Smith MJ. Cryopreservation of zebra fish spermatozoa using methanol. Canadian Journal of Zoology. 1982;60(8):1867-70.

[53] Cosson MP, Billard R, Gatti JL, Christen R. Rapid and quantitative assessment of trout spermatozoa motility using stroboscopy. Aquaculture. 1985;46(1):71-5.

[54] Billard R, Cosson MP. Some problems related to the assessment of sperm motility in freshwater fish. Journal of Experimental Zoology. 1992;261(2):122-31.

[55] Ohta H, Shimma H, Hirose K. Relationship between fertility and motility of cryopreserved spermatozoa of the amago salmon *Oncorhynchus masou ishikawae*. Fisheries science. 1995;61(5):886-7.

[56] Benau D, Terner C. Initiation, prolongation, and reactivation of the motility of salmonid spermatozoa. Gamete Research. 1980;3(3):247-57.

[57] Akcay E, Bukan N, Bozkurt Y, Tekin N, Secer S. Correlation between biochemical and spermatological parameters in rainbow trout (*Oncorhynchus mykiss*) semen. Israeli Journal of Aquaculture-Bamidgeh. 2004;56:20390.

[58] Babiak I, Ottesen O, Rudolfsen G, Johnsen S. Quantitative characteristics of Atlantic halibut, *Hippoglossus hippoglossus* L., semen throughout the reproductive season. Theriogenology. 2006;65(8):1587-604.

[59] Mochida K, Kondo T, Matsubara T, Adachi S, Yamauchi K. A high molecular weight glycoprotein in seminal plasma is a sperm immobilizing factor in the teleost Nile tilapia, *Oreochromis niloticus*. Development, growth & differentiation. 1999;41(5):619-27.

[60] Cherr GN, Morisawa M, Vines CA, Yoshida Y, Smith EH, Matsubara T, Pillai M, Griffin FJ, Yanagimachi R. Two egg-derived molecules in sperm motility initiation and fertilization in the Pacific herring (*Clupea pallasi*). University of the Basque Country Press, 52, pp.743-752.

[61] Morisawa M, Suzuki K. Osmolality and potassium ion: their roles in initiation of sperm motility in teleosts. Science. 1980;210(4474):1145-7.

[62] Billard R, Weil C, Bieniarz K, Mikolajczyk T, Breton B, Epler P, Bougoussa M. Testicular and some hormonal changes during the first four

years of life in the mirror carp, *Cyprinus carpio* L. Journal of fish biology. 1992;41(3):473-87.

[63] Perchec G, Jeulin C, Cosson J, Andre F, Billard R. Relationship between sperm ATP content and motility of carp spermatozoa. Journal of Cell Science. 1995;108(2):747-53.

[64] Cosson J. Fish sperm physiology: structure, factors regulating motility, and motility evaluation. Biological Research in Aquatic Science. 2019;1: 1-26.

[65] Verma DK, Routray P, Dash C, Dasgupta S, Jk JE. Physical and biochemical characteristics of semen and ultrastructure of spermatozoa in six carp species. Turkish Journal of Fisheries and Aquatic Sciences. 2009;9(1):pp.67-76

[66] Christen R, Schackmann RW, Shapiro BM. Metabolism of sea urchin sperm. Interrelationships between intracellular pH, ATPase activity, and mitochondrial respiration. Journal of Biological Chemistry. 1983;258(9): 5392-5399.

[67] Shapiro BM, Schackmann RW. Coupled ionic and enzymatic regulation of sperm behavior. Modulation by Molecular Interactions: Current Topics in Cellular Regulation, Vol. 26. 2013;26:97-113.

[68] Márián TE, Krasznai ZO, Balkay L, Balazs M, Emri M, Bene L, Tron L. Hypo-osmotic shock induces an osmolality-dependent permeabilization and structural changes in the membrane of carp sperm. Journal of Histochemistry & Cytochemistry. 1993;41(2):291-297.

[69] Boitano S, Omoto CK. Trout sperm swimming patterns and role of intracellular Ca++. Cell motility and the cytoskeleton. 1992;21(1):74-82.

[70] Agarwal NK. Developing cryopreservation protocols for the spermatozoa of snow trout and other endangered coldwater species. [Final report]. 2005.

[71] Bozkurt Y, Öğretmen F. Sperm quality, egg size, fecundity and their relationships with fertilization rate of grass carp (*Ctenopharyngodon idella*). Iranian Journal of Fisheries Sciences. 2012;11(4):755-64.

[72] Kruger JD, Smit GL, Van Vuren JH, Ferreira JT. Some chemical and physical characteristics of the semen of *Cyprinus carpio* L. and *Oreochromis mossambicus* (Peters). Journal of Fish Biology. 1984;24(3):263-272.

[73] Suquet M, Dreanno C, Fauvel C, Cosson J, Billard R. Cryopreservation of sperm in marine fish. Aquaculture Research: Original Articles. 2000;31(3):231-43.

[74] Bozkurt Y, Öğretmen F, Kökçü Ö, Ercin U. Relationships between seminal plasma composition and sperm quality parameters of the *Salmo trutta macrostigma* (Dumeril, 1858) semen: with emphasis on sperm motility. Czech J. Anim. Sci. 2011;56(8):355-64.

[75] Routray P, Verma DK. Utilization of milt cryopreservation technology in aquaculture-a strategy. In *National Workshop On "Portable FRP Carp Hatchery Technology"*, 2011:11, pp.28.

[76] Graybill JR, Horton HF. Limited fertilization of chinook and coho salmon eggs with cryopreserved sperm. J. Fish. Res. Board Can. 1969;26(5):1400-4.

[77] Kamaruding NA, Embong WK, Abdullan RB. Frozen-thawed sperm motility characteristics of African catfish (*Clarias gariepinus*) by using glycerol or DMSO based extender. International Journal of Environmental Science and Development. 2012;3(1):49.

[78] Borchard B, Schmidt GW. Versuche mit Regenbogenforellensperma. IV. Die Tiefkuhlkonservierung. Beobachtungen zum Einsatz bei praktischen und wissenschaftlichen Arbeiten, Fischwirt. 1979;29:49-51.

[79] Ohta H, Izawa T. Diluent for cool storage of the Japanese eel (*Anguilla japonica*) spermatozoa. Aquaculture. 1996;142(1-2):107-18.

[80] Bates MC, Wayman WR, Tiersch TR. Effect of osmotic pressure on the activation and storage of channel catfish sperm. Transactions of the American Fisheries Society. 1996;125(5):798-802.

[81] Urbanyi B, Horvath A, Varga Z, Horvath L, Magyary I, Radics F. Effect of extenders on sperm cryopreservation of African catfish, *Clarias gariepinus* (Burchell). Aquaculture Research. 1999;30(2):145-51.

[82] Lakra WS, Krishna G. Preliminary trials for cryopreservation of spermatozoa of selected carps and catfishes. Indian journal of animal sciences. 1997;67(1):90-2.

[83] Irawan H, Vuthiphandchai V, Nimrat S. The effect of extenders, cryoprotectants and cryopreservation methods on common carp (*Cyprinus carpio*) sperm. Animal Reproduction Science. 2010;122(3-4):236-43.

[84] Betsy CJ, Kumar JS. Effect of three different extenders on the spermatological parameters of *Cyprinus carpio* spermatozoa. Journal of Aquaculture in the Tropics. 2016;31(1/2):25.

[85] Cloud J, Patton S. Basic principles of fish spermatozoa cryopreservation. Methods in. 2008:237..

[86] Tekin N, Secer S, Akcay E, Bozkurt Y, Kayam S. Effects of glycerol additions on post-thaw fertility of frozen rainbow trout sperm, with an emphasis on interaction between extender and cryoprotectant. Journal of Applied Ichthyology. 2007;23(1):60-3.

[87] Yavaş İ, Bozkurt Y, Yıldız C. Cryopreservation of scaly carp (*Cyprinus carpio*) sperm: effect of different cryoprotectant concentrations on post-thaw motility, fertilization and hatching success of embryos. Aquaculture international. 2014;22(1):141-8.

[88] Tiersch TR, Yang H, Jenkins JA, Dong Q. Sperm cryopreservation in fish and shellfish. Society of Reproduction and Fertility supplement. 2007;65:493.

[89] Leung LK. Principles of biological cryopreservation. Fish evolution and systematics: evidence from spermatozoa. 1991:231-44.

[90] Meryman HT. Cryoprotective agents. Cryobiology. 1971 Apr 1;8(2):173-83.

[91] Taylor R, Adams GD, Boardman CF, Wallis RG. Cryoprotection—Permeant vs nonpermeant additives. Cryobiology. 1974;11(5):430-8.

[92] Chao NH. Fish sperm cryopreservation in Taiwan: Technology advancement and extension efforts. *Bulletin of the Institute of Zoology, Academia Sinica*, 1991;16:263-83.

[93] Simione F. American Type Culture Collection in cooperation with NALGE Nunc International Corp. Cryopreservation manual. 1998.

[94] Routray P, Dash SN, Dash C, Swain P, Sarkar SK, Sarangi N. Cryopreservation of silver barb *Puntius gonionotus* (Bleeker) spermatozoa: effect of extender composition, cryoprotective agents and freezing rate on their postthawing fertilization ability. Aquaculture Research. 2008; 39(15):1597-605.

[95] Richardson GF, Wilson CE, Crim LW, Yao Z. Cryopreservation of yellowtail flounder (*Pleuronectes ferrugineus*) semen in large straws. Aquaculture. 1999;174(1-2):89-94.

[96] Horvath A, Urbanyi B. The effect of cryoprotectants on the motility and fertilizing capacity of cryopreserved African catfish *Clarias gariepinus* (Burchell 1822) sperm. Aquaculture Research. 2000;31(3):317-24.

[97] Ohta H, Kawamura K, Unuma T, Takegoshi Y. Cryopreservation of the sperm of the Japanese bitterling. Journal of Fish Biology. 2001;58(3):670-81.

[98] Muchlisin ZA, Muhammadar. Long-term cryopreservation of baung spermatozoa, Mystus nemurus: effect of various cryoprotectants on motility and fertility. Torani. 2002;12:204-10.

[99] Viveiros AT, So N, Komen J. Sperm cryopreservation of African catfish, *Clarias gariepinus*: cryoprotectants, freezing rates and sperm: egg dilution ratio. Theriogenology. 2000;54(9):1395-408.

[100] Lahnsteiner F, Berger B, Horvath A, Urbányi B, Weismann T. Cryopreservation of spermatozoa in cyprinid fishes. Theriogenology. 2000;54(9):1477-98.

[101] Kwantong S, Bart AN. Cryopreservation of black ear catfish, *Pangasius larnaudii*, (Bocourt) sperm. Aquaculture Research. 2006;37(9):955-7.

[102] Chao NH, Liao IC. Cryopreservation of finfish and shellfish gametes and embryos. Reproductive Biotechnology in Finfish Aquaculture. 2001:161-89.

[103] Vuthiphandchai V, Chomphuthawach S, Nimrat S. Cryopreservation of red snapper (*Lutjanus argentimaculatus*) sperm: effect of cryoprotectants and cooling rates on sperm motility, sperm viability, and fertilization capacity. Theriogenology. 2009;72(1):129-38.

[104] Betsy CJ, Kumar JS. Effect of egg yolk on enhancing quality of mrigal (*Cirrhinus mrigala*) spermatozoa during cryopreservation. Indian Journal of Fisheries. 2015;62(4):140-3.

[105] Betsy CJ, Kumar JS, Rajagopalasamy CB. Role of glucose in enhancing life and potency of *Cirrhinus mrigala* spermatozoa during cryopreservation. Journal of Applied and Natural Science. 2015;7(1):159-64.

[106] Betsy CJ, Kumar JS. Influence of bovine serum albumin (BSA) on the spermatological parameters of *Cyprinus carpio* during cryopreservation. Journal of Aquaculture in the Tropics. 2016;31(1/2):103.

[107] Billard R. Artificial insemination in salmonids. In: R.N. Iwamoto and S. Sower (Editors), Salmonid Reproduction. Washington Sea Grant Publication.Seattle, WA, 1985:pp. 116- 128

[108] Rieniets JP, Millard JL. Use of saline solutions to improve fertilization of northern pike eggs. The Progressive Fish-Culturist. 1987;49(2):117-9.

[109] McPartlin LA, Littell J, Mark E, Nelson JL, Travis AJ, Bedford-Guaus SJ. A defined medium supports changes consistent with capacitation in stallion sperm, as evidenced by increases in protein tyrosine phosphorylation and high rates of acrosomal exocytosis. Theriogenology. 2008;69(5):639-50.

[110] Lahnsteiner F, Berger B, Weismann T. Effects of media, fertilization technique, extender, straw volume, and sperm to egg ratio on hatchability of cyprinid embryos, using cryopreserved semen. Theriogenology. 2003;60(5):829-41.

[111] Linhart O, Rodina M, Cosson J. Cryopreservation of sperm in common carp *Cyprinus carpio*: sperm motility and hatching success of embryos. Cryobiology. 2000;41(3):241-50.

[112] Babiak I, Brzuska E, Perkowski J. Fractional factorial design of screening experiments on cryopreservation of fish sperm. Aquaculture research. 2000;31(3):273-82.

[113] Magyary I, Urbanyi B, Horvath L. Cryopreservation of common carp (*Cyprinus carpio* L.) sperm II. Optimal conditions for fertilization. Journal of Applied Ichthyology. 1996;12(2):117-9.

[114] Linhart O, Rodina M. Cryopreservation of common carp *Cyprinus carpio* and tench *Tinca tinca* sperm for gene resources conservation. In International Symposium on the Reproductive Physiology of Fish 2000.

[115] Alavi SM, Rodina M, Policar T, Kozak P, Psenicka M, Linhart O. Semen of *Perca fluviatilis* L.: Sperm volume and density, seminal plasma indices and effects of dilution ratio, ions and osmolality on sperm motility. Theriogenology. 2007;68(2):276-83.

[116] Betsy CJ, Kumar JS, Jawahar KT. Effect of different extenders and dilution ratios on the spermatological and fertilization parameters of *Cyprinus carpio* during cryopreservation. Biopreservation and biobanking. 2019;17(2):157-62.

[117] Grout BW, Morris GJ. Effects of low temperatures on biological systems. Edward Arnold; 1986.

[118] Gupta SD, Rath SC. Cryogenic preservation of carp milt and its utilization in seed production. In Proc. 3rd Ind. Fish. Forum 1993:pp. 11-14.

[119] Ponniah AG, Lal KK, Gopalakrishnan A, Srivastava SK. Use of fertilisation protocols to enhance hatching percentage with cryopreserved milt of *Cyprinus carpio.* National Academy Science Letters. 1998;21(7-8):256-60.

[120] Ponniah AG, Sahoo PK, Dayal R, Barat A. Cryopreservation of *Tor putitora* spermatozoa: Effect of extender composition, activating solution, cryoprotectant and equilibration time. *Proceedings-national academy of sciences India section B*. 1999;69(1):53-60.

[121] Rafiquzzaman SM, Sarder MR, Islam MS, Sultana R. Sperm cryopreservation of Indian major carp, *Labeo rohita*: cryodiluents, sperm: cryodiluent dilution ratio and cryoprotectant concentration. Bangladesh Journal of Fisheries Research. 2007;11(2):141-52.

[122] Pérez-Cerezales S, Martínez-Páramo S, Beirão J, Herráez MP. Evaluation of DNA damage as a quality marker for rainbow trout sperm cryopreservation and use of LDL as cryoprotectant. Theriogenology. 2010;74(2):282-9.

[123] Stoss J, Holtz W. Successful storage of chilled rainbow trout (*Salmo gairdneri*) spermatozoa for up to 34 days. Aquaculture. 1983;31(2-4):269-74.

[124] Linhart O, Billard R, Proteau JP. Cryopreservation of European catfish (*Silurus glanis* L.) spermatozoa. Aquaculture. 1993;115(3-4):347-59.

[125] Franks F. *Biophysics and biochemistry at low temperatures*. Cambridge University Press, Cambridge, 1985: pp. 37-61.

[126] Hu E, Yang H, Tiersch TR. High-throughput cryopreservation of spermatozoa of blue catfish (*Ictalurus furcatus*): establishment of an approach for commercial-scale processing. Cryobiology. 2011;62(1):74-82.

[127] Watson PF. The causes of reduced fertility with cryopreserved semen.

Animal reproduction science. 2000; 60:481-92.

[128] Young JA, Capra MF, Blackshaw AW. Cryopreservation of summer whiting (*Sillago ciliata*) spermatozoa. Aquaculture. 1992;102(1-2):155-60.

[129] Kurokura H, Hirano R, Tomita M, Iwahashi M. Cryopreservation of carp sperm. Aquaculture. 1984;37(3):267-73.

[130] Scheerer PD, Thorgaard GH. Improved fertilization by cryopreserved rainbow trout semen treated with theophylline. The Progressive Fish-Culturist. 1989;51(3):179-82.

[131] Ciereszko A, Dabrowski K. Effect of a sucrose–DMSO extender supplemented with pentoxifylline or blood plasma on fertilizing ability of cryopreserved rainbow trout spermatozoa. The Progressive fish-culturist. 1996;58(2):143-5.

[132] Ritar AJ. Artificial insemination with cryopreserved semen from striped trumpeter (*Latris lineata*). Aquaculture. 1999;180(1-2):177-87.

[133] Tiersch TR, Goudie CA, Carmichael GJ. Cryopreservation of channel catfish sperm: storage in cryoprotectants, fertilization trials, and growth of channel catfish produced with cryopreserved sperm. Transactions of the American Fisheries Society. 1994;123(4):580-6.

[134] Horváth Á, Miskolczi E, Urbányi B. Cryopreservation of common carp sperm. Aquatic Living Resources. 2003;16(5):457-60.

[135] Cognie PF, Billard R, Chao NH. La cryoconservation de la laitance de la carpe, *Cyprinus carpio*. Journal of applied ichthyology,1989; 5(4), pp.165-176.

[136] Sultana M, Nahiduzzaman M, Hassan MM, Khanam MU, Hossain MA. Fertility of cryopreserved common carp (*Cyprinus carpio*) spermatozoa. University Journal of Zoology, Rajshahi University. 2009;28:51-5.

[137] Vuthiphandchai V, Wilairattanadilok K, Chomphuthawach S, Sooksawat T, Nimrat S. Sperm cryopreservation of silver barb (*Barbodes gonionotus*): cryoprotectants, cooling rate and storage time on sperm quality. Aquaculture Research. 2015;46(10):2443-51.

[138] Gorin NC. Cryopreservation and storage of stem cell. Bone Marrow and Stem Cell Processing.1992;pp.292-308.

[139] Rana KJ, McAndrew BJ. The viability of cryopreserved tilapia spermatozoa. Aquaculture. 1989; 76(3-4):335-45.

[140] Linhart O, Billard R, Proteau JP. Cryopreservation of European catfish (*Silurus glanis* L.) spermatozoa. Aquaculture. 1993;115(3-4):347-59.

[141] Conget P, Fernández M, Herrera G, Minguell J. Cryopreservation of rainbow trout (*Oncorhynchus mykiss*) spermatozoa using programmable freezing. Aquaculture. 1996;143(3-4):319-29.

[142] Warnecke D, Pluta HJ. Motility and fertilizing capacity of frozen/thawed common carp (*Cyprinus carpio* L.) sperm using dimethyl-acetamide as the main cryoprotectant. Aquaculture. 2003;215(1-4):167-85.

[143] Lahnsteiner FR. Semen cryopreservation in the Salmonidae and in the Northern pike. Aquaculture Research. 2000;31(3):245-58.

[144] Richardson GF, McNiven MA, Mansour N. Effect of methanol concentration and thaw rate on the viability and fertility of cryopreserved Arctic char, *Salvelinus alpinus* (L.),

spermatozoa. Aquaculture Research. 2011;42(8):1096-100.

[145] Durbin H, Durbin FJ, Stott B. A note on the cryopreservation of grass carp milt. Aquaculture Research. 1982;13(3):115-7.

[146] Withler FC. Cryopreservation of spermatozoa of some freshwater fishes cultured in South and Southeast Asia. Aquaculture. 1982;26(3-4):395-8.

[147] Kumar K. A comparative-study of various extenders for cryopreservation of carp spermatozoa. Indian Journal of Animal Sciences. 1988; 58(11): 1355-1360.

[148] Basavaraja N, Hegde SN. Cryopreservation of the endangered mahseer (*Tor khudree*) spermatozoa: I. Effect of extender composition, cryoprotectants, dilution ratio, and storage period on post-thaw viability. Cryobiology. 2004;49(2):149-56.

[149] Yavas I, Bozkurt Y. Effect of different thawing rates on motility and fertilizing capacity of cryopreserved grass carp (*Ctenopharyngodon idella*) sperm. Biotechnology & Biotechnological Equipment. 2011;25(1):2254-7.

[150] Bozkurt Y, Yavaş İ. Effect of extender compositions, glycerol levels, and thawing rates on motility and fertility of cryopreserved wild African Catfish (*Clarias gariepinus*) sperm. The Israeli Journal of Aquaculture.2017;69:1357-1364.

[151] Lahnsteiner F, Weismann T, Patzner RA. Methanol as cryoprotectant and the suitability of 1.2 ml and 5 ml straws for cryopreservation of semen from salmonid fishes. Aquaculture Research. 1997;28(6):471-9.

[152] Zell SR. Cryopreservation of gametes and embryos of salmonids fishes. Ann. Biol. Anim. Biochem. Biophys. 1978;18: 1089-1099.

[153] Erdahl DA, Graham EF. Preservation of gametes of freshwater fish, in Proceedings of International Conference of Animal Reproduction and Artificial Insemination, Madrid. 1980:317-326.

[154] Arii N, Namai K, Gomi F, Nakazawa T. Cryoprotection of medaka embryos during development. Zoological science. 1987;4(5):813-8.

[155] Nilsson E, Cloud JG. Cryopreservation of rainbow trout (*Oncorhynchus mykiss*) blastomeres. Aquatic Living Resources. 1993;6(1):77-80. doi: 10.1006/cryo.1998.2084

[156] Zhang T, Rawson DM, Morris GJ. Cryopreservation of pre-hatch embryos of zebrafish (*Brachydanio rerio*). Aquatic Living Resources. 1993;6(2):145-53.

[157] Ahammad MM, Bhattacharyya D, Jana BB. Stage-dependent hatching responses of rohu (*Labeo rohita*) embryos to different concentrations of cryoprotectants and temperatures. Cryobiology. 2003;46(1):2-16. doi: 10.1016/s0011-2240(02)00138-4

[158] Harvey B, Kelley RN, Ashwood-Smith MJ. Permeability of intact and dechorionated zebra fish embryos to glycerol and dimethyl sulfoxide. Cryobiology. 1983;20(4):432-9.

[159] Horton HF, Ott AG. Cryopreservation of fish spermatozoa and ova. Journal of the Fisheries Board of Canada. 1976;33(4):995-1000.

[160] Loeffler CA, Løvtrup S. Water balance in the salmon egg. Journal of experimental Biology. 2000;52(2):291-8.

[161] Diwan AD, Harke SN, Panche AN. Cryobanking of Fish and Shellfish Egg, Embryos and Larvae: An Overview. Frontiers in Marine Science. 2020;7:251. doi: 10.3389/fmars.2020.00251

[162] Rall WF. Advances in the cryopreservation of embryos and

prospects for application to the conservation of salmonid fishes. In Genetic conservation of salmonid fishes. Springer, Boston, MA. 1993:pp. 137-158

[163] Jamieson BG, Leung LP. Fish evolution and systematics: evidence from spermatozoa: with a survey of lophophorate, echinoderm and protochordate sperm and an account of gamete cryopreservation. Cambridge University Press; 1991.

[164] Routray P, Suzuki T, Strüssmann CA, Takai R. Factors affecting the uptake of DMSO by the eggs and embryos of medaka, *Oryzias latipes*. Theriogenology. 2002;58(8): 1483-96. doi: 10.1016/s0093-691x(02) 01076-2

[165] Robles V, Cabrita E, Anel L, Herráez MP. Microinjection of the antifreeze protein type III (AFPIII) in turbot (*Scophthalmus maximus*) embryos: toxicity and protein distribution. Aquaculture. 2006;261(4): 1299-306.

[166] Valdez Jr DM, Miyamoto A, Hara T, Edashige K, Kasai M. Sensitivity to chilling of medaka (*Oryzias latipes)* embryos at various developmental stages. Theriogenology. 2005;64(1):112-22. doi: 10.1016/j.theriogenology.2004. 11.006

[167] Streit Jr DP, Godoy LD, Ribeiro RP, Fornari DC, Digmayer M, Zhang T. Cryopreservation of embryos and oocytes of South American fish species. In Tech; 2014.

[168] Whittingham DG, DG W. Attempts to preserve herring embryos at subzero temperatures. Arc Fischeriwiss 1978; 29: 75-9.

[169] Zhang T, Rawson DM, Morris BJ. Cryopreservation of prehatch embryo of zebrafish. Aquatic Living Resources. 1993; 6: 145-53.

[170] Hagedorn M, Hsu E, Kleinhans FW, Wildt DE. New approaches for studying the permeability of fish embryos: toward successful cryopreservation. Cryobiology. 1997;34(4):335-47.

[171] Anchordoguy TJ, Rudolph AS, Carpenter JF, Crowe JH. Modes of interaction of cryoprotectants with membrane phospholipids during freezing. Cryobiology. 1987;24(4):324-31.

Section 4

Reproduction in Avian Species

Chapter 6

Avian Reproduction

Kingsley Omogiade Idahor

Abstract

There are about 10,400 living avian species belonging to the class Aves, characterized by feathers which no other animal classes possess and are warm-blooded vertebrates with four-chamber heart. They have excellent vision, and their forelimbs are modified into wings for flight or swimming, though not all can fly or swim. They lay hard-shelled eggs which are a secretory product of the reproductive system that vary greatly in colour, shape and size, and the bigger the bird, the bigger the egg. Since domestication, avian species have been basically reared for eggs, meat, pleasure and research. They reproduce sexually with the spermatozoa being homogametic and carry Z-bearing chromosomes, and the blastodisk carries either Z-bearing or W-bearing chromosomes, hence, the female is heterogametic, and thus, determines the sex of the offspring. The paired testes produce spermatozoa, sex hormones and the single ovary (with a few exceptions) produces yolk bearing the blastodisk and sex hormones. Both testis and ovary are the primary sex organs involved in sexual characteristics development in avian. In avian reproduction, there must be mating for fertile egg that must be incubated to produce the young ones. At hatch, hatchling sex is identified and reared to meet the aim of the farmer.

Keywords: avian species, chromosomes, embryogenesis, hatchability, incubation

1. Introduction

1.1 Avian species

Avian species is a group of warm-blooded vertebrates constituting the class Aves, characterized by feathers, toothless beak, laying of hard-shelled eggs, high metabolic rate, four-chambered heart and a strong but lightweight skeleton. They vary in size from about 5.5 cm long (i.e. bee hummingbird) to as long as 2.8 m (i.e. ostrich). There are about 10,400 living avian species spread widely across the world with more than half capable of flying (i.e. passerine or perching birds). They have wings that vary widely depending on the species; however, there existed the Moa and Elephant birds that were the only known groups without wings. The wings normally develop from the forelimbs for flight and the tail feathers for flight control. Meanwhile, further evolvement has led to loss of flight ability in some birds such as ratites, penguins and diverse endemic island species even with well-developed feathers and wings. However, the digestive and respiratory systems of birds are physiologically suited for flight; hence, some avian species have adapted to aquatic environments for survival such as seabirds and water birds; thus, they have the capability to swim and feed in water [1].

Birds are feathered theropod dinosaurs and constitute the only living dinosaurs yet, could be considered as reptiles in the modern cladistics viewpoint, and their closest living relatives are the Crocodilians. They are believed to be descendants of the primitive Avialans (whose members include Archaeopteryx that first appeared about 160 million years ago in China [2]. According to deoxyribonucleic acid evidence, modern birds categorized as Neornithes evolved in the middle to late Cretaceous and diversified tremendously around the time with Paleogene extinction event about 66 million years ago that exterminated the Pterosaurs and all non-avian dinosaurs [3]. Several of the social bird species pass on knowledge from generation to generation because of their ability to communicate with visual signals, calls, sing and participate in such behaviours as cooperative breeding and hunting, flocking and mobbing of predators. These attributes could best explain why birds are sometimes referred to as social and cultural animals. Most bird species are socially monogamous but not necessarily sexually, because this habit may be for a breeding season, sometimes for years but rarely for life. Some other species have breeding systems that are polygamous, i.e. one male with several females and seldom polyandrous, where a female has various males. Birds produce their progenies by sexual reproduction, where the female releases yolk with a blastodisk on the surface that must be fertilized by the spermatozoon released by the male during mating or artificial insemination. The female usually lays fertilized eggs in a nest and incubated for a length of time characterized by the avian species. Although the hatchlings could be altricial or precocial, the dam/dams (i.e. parent/parents) usually have an extended period of parental care after hatching except, a few species such as Australian brush turkeys that do not exhibit this habit [4].

There are more than 10,400 extant bird species in the world. Across North America and South America alone, there are more than 4400 species, approximately 2700 different species in Asia, and about 2300 are found in Africa. In Europe (west of the Ural Mountains), there are more than 500 species, and more than 700 species are found in Russia with Costa having the highest concentrations of roughly 800 bird species.

1.2 Determination of sex in avian species

According to Osinowo [5], the spermatozoa in avian species are homogametic and carry Z-bearing chromosome, whereas the ovum (blastodisk/blastoderm) carries either Z-bearing or W-bearing chromosomes. Therefore, the female is heterogametic and as such determines the sex of the offspring. At fertilization, the union of Z-bearing chromosomes from both the male and female will result in male offspring (ZZ) while the union of a W-bearing chromosome ovum (blastoderm) develops into female offspring (ZW). Consequently, determination of sex mechanisms in avian species differs from that of the mammals where the male is the sex determinant. However, aberrations may occur due to non-disjunction of sex chromosomes, translocation, deletion or mutation of genes forming ZZZ, ZZW or ZZO zygotes with attendant defects as recorded at the Department of Animal Science, University of Ibadan, Nigeria [6].

1.3 Male reproductive system

In all avian species, the paired testes which are the gonads that produce the gametes—spermatozoa are retained in the abdominal cavity towards the cephalic border of the kidneys, where they originated from the mesonephric ducts forming the Wolfian duct. Therefore, the testes have the same temperature (40°C) as the body yet, spermatogenesis occurs, and the spermatozoa remain viable at body temperature. Unlike mammals that require scrotum for thermoregulation before spermatogenesis can occur and be viable. According to Johnson [7], each of the testes is attached to the body wall by the mesorchium and is encapsulated by a fibrous inner coat, the tunica albuginea and a thin outer layer, the tunica vaginalis. One of the two testes may be larger (depending on the species) yet, both will be functional. The weight of the testes in chickens constitutes about 1% of the total body weight, depending on the breed and about 9–30 g per testis at sexual maturity. However, in seasonal breeders, testis size may increase by 300–500 folds during the reproductive active season as compared with the nonbreeding state.

Although many environmental factors impact on reproductive activity, the seasonal breeder responds most strongly to long day length. Testicular interstitial cells produce testosterone that influences reproductive behaviour such as territorial aggression and song. Other changes observed in seasonal breeders include testicular hypertrophy and enlargement of the ductus deferens and seminal glomus. The testis is elliptical, yellowish and consists of a large number of slender and convoluted ducts called seminiferous tubule where spermatozoa are produced. These ducts connect to the paired vas deferens terminating in the small papilla that serves as an intromittent or a copulatory organ. Although birds are one of the groups which reproduce through internal fertilization, they have repeatedly lost the intromittent organ; thus, most avian species do not have penis. However, larger birds such as duck, ostrich and emu have penis. Ostrich has a conical-shaped penis that is wider at the base as given by Brennan and Prum [8]. Even if birds reproduce through internal fertilization, 97% of the males absolutely lack a functional intromittent organ. While the other 3% have intromittent organ, copulation occurs through brief insertion of the male organ into the vagina before ejaculation. On the other hand, vast majority of the birds comprising nearly 10,000 species transfer spermatozoa via cloacal contact between the male and female in a manoeuvre described as 'cloacal kiss' [9]. Nevertheless, a functional intromittent organ is known to be present in most species of Palaeognathae (ostriches, rheas, kiwis, tinamous, cassowaries and emu) and Anseriformes (waterfowl, ducks, geese and swans) with high variability in the intromittent organ morphology.

In waterfowl, for example, the intromittent organ varies greatly in length, characterized by surface elaborations (both spines and grooves) and spiral counterclockwise [10]. This variation is most likely due to an intersexual arms race resulting from a mating system in which forced extra-pair copulations are frequent. While drake has a penis that is coiled along the ventral wall of the cloaca when flaccid, it has an elaborate spiral shape when erect. The drakes often force sex on the ducks to scatter their genes, and the ducks have evolved complex anatomical defenses against these unwanted attentions. But the lymphatic erection in the male offers a way around these, because it allows a rapid on/off means of extending the penis to enable deep insemination [11].

1.3.1 Spermatogenesis

The sequence of events in the development of avian spermatozoa from spermatogonia is known as spermatogenesis. It occurs in the seminiferous tubules in the testes and involves several physiological processes such as spermatocytogenesis, meiosis and spermeogenesis [12]. The germ/stem/primitive/primordial cells or the spermatogonia develop into primary and secondary spermatocytes. During the meiotic phase, maturation division of the spermatocytes results in spermatids with a reduced number of chromosomes and gradually transforms into spermatozoa. In some avian species such as guinea fowl and domestic fowl, spermeogenesis involves 10 different morphological sequences but about 12 in Japanese quails [13–15]. During spermeogenesis, biochemical processes involving acrosome and axoneme formation, loss of cytoplasm and replacement of nucleohistones with nucleoprotamine accompanied with nuclei condensation and transformation of spermatids into spermatozoa without further cell division occur [16].

Consequently, spermatogenesis sequences could be simplified as Spermatogonia → primary spermatocytes → secondary spermatocytes → spermatids → spermatozoa. The produced spermatozoa are stored in the epididymis and vas deferens until ejaculation, when they will be released through the intromittent organ. The semen volume in avian species is probably the lowest among livestock species but has the highest spermatozoa concentration per semen volume. Adeoye et al. [17] and Almahdi et al. [18] attributed this to lack of accessory glands in avian species which are well developed in other livestock species particularly mammals.

1.4 Female reproductive system

Essentially, avian species do not have cervix in their reproductive system, thus the lower end of the oviduct opens into the cloaca. The cloaca contains openings for the reproductive, digestive and urinary tracts. The paired ovaries are only present in most avian species at few days old (about 5–10 days of post-hatch). Thereafter, the right ovary and the right fallopian tube regress, leaving only the left ovary and the left fallopian tube near the kidneys, where they first differentiated from the paramesonephric duct forming the Müllerian duct. However, Fitzpatrick [19] reported that red-tailed hawk (*Buteo b. borealis*) has clearly a definite vestige of a right ovary consisting of 23 follicles at adult stage. If with a single ovary, some poultry species have the capability to lay several eggs per annum (200–350 eggs in chicken), what if both ovaries were intact and functional? The physiological explanation to this phenomenon seems not to be available, because some avian species such as hawks and owls have both ovaries intact [20] yet, they can only lay a few eggs per annum (about 2–5 eggs per clutch in a reproduction cycle). At hatch, avian female hatchling's ovary has about 4000 follicles and will never have new ova produced in life. Out of this, only some are fully developed and ovulated in a life span of a mature poultry female bird. The large follicles are round, yellow and loosely connected or attached to the ovary by the follicular stalk, and the yolk sac is richly supplied with blood and nutritional materials. The largest follicle containing the blastodisk ovulates when the yolk sac ruptures. Only one large follicle is ovulated at a time but two or more may be ovulated resulting in double yolk egg or seldom double eggs with a shelled egg inside another shelled egg. **Figure 1** shows a typical female avian reproductive tract as culled from Suyatma and Hermanianto [21].

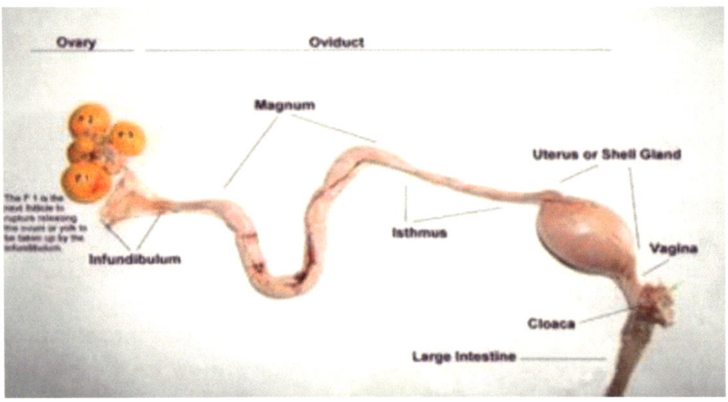

Figure 1.
Female avian reproductive tract. Source: Suyatma and Hermanianto [21].

Avian oviduct is a complex organ with different segments that converts nutrients from the feed consumed into the various components of a well-formed egg. Early changes associated with rising oestrogen levels in female, include osteomyelosclerosis and hypercalcemia. Ovulation is then induced by leutinizing hormone, which is followed by eggshell calcification under the control of progesterone. Sources of calcium for shell production include intestinal absorption from the diet, renal control of calcium levels and mobilization of bone calcium deposits. During oviposition, PGF2 alpha and vasotocin stimulate powerful uterine contractions in the presence of calcium. Incubation is associated with falling leutinizing hormone levels and rising prolactin levels. If the hen actually enters reproductive quiescence at this time, then molt will follow. Molt is associated with the total regression of the reproductive tract [22].

1.5 Avian egg formation

Avian egg formation is independent of mating because once mature yolk with the ovum (blastodisk) is released, it is swept into the fallopian tube and stays for a short time in the ampulla region where fertilization occurs. With fertilization or no fertilization, chalaza is secreted on the yolk within 15 min in the infundibulum and then moves to the magnum which is the longest portion of the oviduct and albumen is secreted within 3 h. Then, it moves to the isthmus where shell membrane is secreted within 1 h 15 min. The whole combinations move to the shell gland portion of the oviduct where calciferous shell is secreted to encapsulate the yolk, chalaza and albumen forming a shelled egg. It takes about 18–22 h in the shell gland before moving to the vagina where cuticle is secreted to cover the shell pores, and it stays for a few minutes in the vagina for mucous secretion that will aid lay. It takes about 26 h for an egg to be fully formed in the avian oviduct before lay; hence, the hen will skip a day after laying for six days (6/7 days). The large follicles, usually about five, are in graded sizes with the largest ovulating first. A mature ovary weighs 40–60 g. All avian species lay eggs but only the poultry birds such as fowl, turkey, duck and quail eggs are commonly available for human consumption. According to Campbell et al. [23], egg lay in birds results from a complex natural endowment whose prime aim is to procreate. Many other animal species including insects, worms, fishes, reptiles and mammals produce eggs, but the avian eggs are much larger than others due to the food reserve for embryonic development. Avian egg is a secretory product of the reproductive system that varies greatly in colour, shape and size, and the bigger the bird, the bigger the egg. Also, the laying capacity of avian birds varies greatly as seen in **Table 1** culled from Campbell et al. [23]. Meanwhile, regular removal of eggs from the nest may increase the rate of egg production among avian

S/N	Avian species	Number of eggs laid per annum
1	Hornbills	1
2	Pigeons	2–4
3	Gulls	4
4	Graylag geese	5–6
5	Mallard ducks	9–11
6	Ostriches	12–15
7	Partridges	12–20
8	Fowl	350 and above

Table 1.
Avian birds laying capacity.

species; however, some birds may abandon the nest, if the eggs are removed, and others may continue to lay in order to establish a clutch for incubation.

1.5.1 Hormonal regulation of avian egg formation

Egg production process is dependent on hormone synchronization and balance. Otherwise, hormone secretion without awaiting the proper signal may result in yolkless, thin-shelled and shell-less eggs as well as formation of shelled egg inside another shelled egg. Essentially, the physical appearance and functioning of avian species could be affected by endocrine secretions. Therefore, some endocrine effects may result from direct action of a single hormone. Hence, the physiological activities of avian species, particularly the female, are dependent on a complex interrelationship of glandular effects as found in the complex hormonal control of ovulation and egg formation [23]. The avian oviduct is usually under control and is stimulated at the most appropriate time to receive the released yolk containing the blastodisk. Ovarian follicle secretions are responsible for the enlargement of the oviduct, vent, spread of the pubic bones, female plumage pattern, mobilization of fat deposit in the yolk and calcium for shell formation. Also, the secretion of albumen is apparently under the control of androgen synthesized by the ovarian interstitial tissue. While eggshell formation is partially controlled by parathyroid glands, the thyroid gland partially controls the seasonal changes in egg laying, feather colouration and feathering during molting.

1.6 Mating

The males mount the females during mating, and the spermatozoa are introduced into the cloaca using intromittent organ. In avian species, mating does not play any role in egg production; however, time of mating determines the rate of egg fertility because it is believed that the eggs may obstruct migration of the spermatozoa to the fallopian tube where fertilization occurs. Spermatozoa are capable of staying up to 3 weeks or 3 months (depending on the avian species) in the uterovaginal portion of the genital organ called 'spermatozoa storage tubule'. Thus, even after withdrawal of the males from the flock or cage or cessation of artificial insemination, the females can still lay fertilized eggs for up to 10–21 days [24]. The uterovaginal junction in the female reproductive part functions as spermatozoa storage tubules. Thus, after a single mating or insemination, the spermatozoa migrate through the vaginal to the spermatozoa storage tubules for subsequent release to fertilize the ovum on the yolk at the ampulla section of the fallopian tube region of the oviduct.

Before now, it was believed that subsequent release of the spermatozoa from the spermatozoa storage tubule was not regulated but occurs in response to the mechanical pressures of a passing ovum, because no contractile elements associated with the spermatozoa storage tubule were found [25, 26]. Recently, several studies reported that spermatozoa maintenance and release from the spermatozoa storage tubules are events regulated during the ovulatory cycle. This was demonstrated by Matsuzaki et al. [27] when progesterone stimulated the release of resident spermatozoa from the spermatozoa storage tubules in Japanese quail with a contraction-like morphological change of the spermatozoa storage tubules. Also, it was shown that the release process was somewhat supported by the lubricant effect of cuticle materials secreted from the ciliated cells of the uterovaginal junction as well as unknown materials supplied from the spermatozoa storage tubules epithelial cells, in events coincidently triggered under progesterone control. Furthermore, Ito et al. [28] found secretory granules in spermatozoa storage tubules epithelial cells and the number of the secretory granules fluctuated during the ovulatory cycle, indicating that spermatozoa storage tubules

S/N	Avian species	Spermatozoa storage duration (weeks)	References
1	Fowl	2–3	Brillard [35]
2	Turkey	10–15	Brillard [35]
3	Quail	1–2	Adeyina et al. [36]

Table 2.
Spermatozoa storage duration of some avian species.

epithelial cells and unknown materials in the lumen of the spermatozoa storage tubules possibly influenced spermatozoa motility, respiration and metabolism.

In some other animal species such as bat, Holt [29] reported that the spermatozoa in the oviduct could be stored for up to 5 months, and Holt and Lloyd [30] stated that reptiles such as turtles, snakes and lizards have obvious potential for spermatozoa storage in the oviduct for an extremely long period of up to 7 years. This phenomenon appears to guarantee and insure against not finding mating partners in some breeding seasons as well as optimizes the timing of the birth of their offspring until a suitable season for nursing arrives. Perhaps, the most remarkable duration of spermatozoa storage was observed in bees [31] and ants [32] that can store spermatozoa for nearly their entire lives. In domestic birds including chickens, turkeys, quails and ducks, Bakst [33] and Bakst et al. [34] stated that once ejaculated spermatozoa enter the female reproductive tract, they can survive for up to 2–15 weeks depending on the species. Spermatozoa storage duration of some avian species is presented in **Table 2**. The disparities may be related to the varying number of spermatozoa storage tubules present in the uterovaginal junction of the avian species (see **Table 3**).

The number of spermatozoa storage tubules in the uterovaginal junction may determine the rate of egg fertility in avian species.

1.7 Fertilization in avian species

If an egg is carefully windowed and the content emptied into a dish, an opaque circular white spot could be seen on the yolk's surface (see **Figure 2**). That spot is called blastodisk in unfertilized or table eggs and blastoderm in fertilized or hatchable eggs. The blastodisk or blastoderm measures about 3–4 mm in diameter in most avian species, and it contains the chromosomes.

The most important part of an egg is the nucleus or germ that develops into the embryo if there where fusion of the pronuclei of the spermatozoon cell and the germinal disk. Other components of the egg provide food and protection for the embryo. An avian egg has shell, shell membrane, albumen, yolk and germinal disk. The eggshell is composed of about 8000 pores for water and gaseous exchange between the egg and the environment. However, a thin film of protein material called cuticle tends to seal the pores in order to reduce loss of water, gases and prevent microbial invasion. Fertilization is the union of the male and female gametes to produce

S/N	Avian spp	No. of spermatozoa storage tubules	References
1	Fowl	5–13,000	[37, 38]
2	Turkey	20–30,000	[37, 38]
3	Quail	3467	[39]

Table 3.
Number of spermatozoa storage tubules found in some avian species.

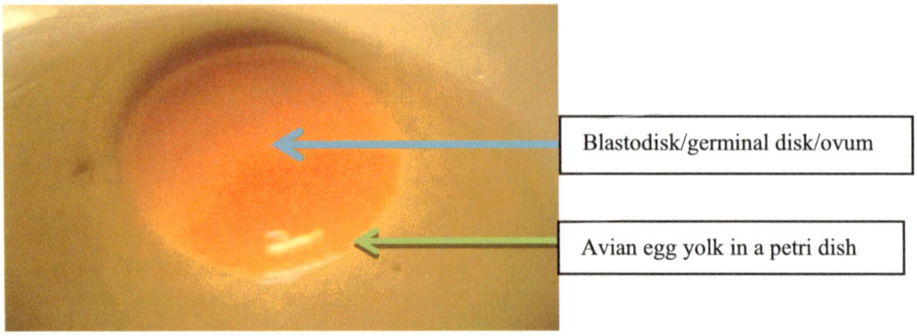

Figure 2.
Avian yolk with blastodisk on the surface.

single-celled zygote. When the yolk is fully mature, it is released from the ovary into the peritoneal cavity where it is swept into the infundibulum and stays in the ampulla region awaiting union with the spermatozoon. It is only in avian species that the ovary releases large yolk with the ovum (also called blastodisk) on the surface.

Following deposition of semen in the uterovaginal region of the oviduct, the spermatozoa migrate to the ampulla section of the fallopian tube to fertilize the egg (blastodisk) on the yolk. Meanwhile, the eggs usually become fertile about four days after the rooster was introduced to the female. However, without mating, insemination and fertilization, avian species still produce shelled eggs. Such eggs are referred to as table eggs or unfertilized eggs or unhatchable eggs; thus, they are purely for human consumption and will never hatch if incubated. But if there was mating, insemination or fertilization, the eggs that will be produced may be fertilized. Such eggs are referred to as fertile or hatchable eggs and could be incubated to produce hatchlings. Avian egg fertility could be influenced by age, nutritional plane, genetic inconsistency, mating or insemination failure and environmental factors. This eventually affects the flock fertility depending on the egg production capability of the female or semen production capability of the male or both sexes. For instance, if the spermatozoa are promptly released from the spermatozoa storage tubules, there might be decrease in egg fertility, and if the male is too old, there might be decrease in spermatozoa quantity and quality resulting in fertility failure. Also, if the male is not interested in mating, there may be irregular mating activity resulting in mating failure. The process of fertilization in avian species involves the release of ovum (blastodisk) that must be swept into the infundibulum and migration of the deposited spermatozoa to meet the ovum at the ampulla region of the fallopian tube where capacitation and syngamy occur.

1.7.1 Capacitation

After ejaculation, the spermatozoa migrate through the uterine body to the ampulla region of the infundibulum. Immediately the spermatozoon comes in contact with the blastodisk on the yolk surface, a physiological mechanism involving enzymatic activity (hyaluronidase) is initiated to remove antigenic seminal factors from the surface of the spermatozoon. This process is referred to as capacitation.

1.7.2 Syngamy

This is a process in which the capacitated spermatozoon crosses the ovum barriers (i.e. cumulus oophorus, zona pellucida, perivitelline and plasma membrane) before final fusion of the pronuclei, resulting in a single-celled zygote. This will subsequently undergo series of mitotic and meiotic division, during cleavage and differentiation into the various parts of the embryo. The embryo derives nutrients from the yolk and albumen, grows rapidly within the eggshell and emerges as hatchling following incubation.

1.8 Avian eggs incubation

Incubation is the process of providing optimum temperature, air circulation and relative humidity suitable for embryo development, growth and emergence as hatchlings. This process could be natural, where the broody hen sits on the eggs and covers them with the feathers in order to provide suitable environmental conditions for hatching. Since some avian species do not naturally incubate and hatch their eggs, artificial incubators are used to simulate environmental conditions required to stimulate embryonic development and growth until the emergence of hatchlings.

1.8.1 Natural incubation

In some avian species, the female birds lay eggs every day or every other day and after laying some eggs that the feathers can cover fully, it stops instinctively and begins to sit on the eggs for incubation. In fowl, for example, once a hen lays a clutch of eggs normally between 3 and 12 eggs, instincts take over.

Thus, it constantly fusses over them, adjusting them just so throughout the day and rarely leaving the nest for more than a few minutes. The broody hen rotates the eggs during incubation for about 96 times in 24 h and keeps the eggs at the correct humidity by splashing water on them from its beak. Motherhood is a big responsibility for a female bird because if it is neglectful, the incubated eggs will never hatch and, if hatched, the hatchlings may be deformed [40]. However, in some other avian species, both male and female are involved in the incubation process. In few occasions, two hens may mutually incubate the eggs that were either laid by one of the hens or both of them (see **Figure 3a–d**).

In these cases, when either of the couple or the two hens is on bouting, the other sits on the eggs. Generally, at the beginning of the incubation, the female sits on the eggs for a longer time before taking a bout. Meanwhile, when ambient temperature is high, the hens spend less time sitting on the eggs, and some hens deliberately cool-off their eggs by sprinkling water or standing in front of the eggs just to provide shadow over the eggs. Whereas in penguin, the female lays an egg and dives into the sea to feed, leaving the male to warm up and incubate the egg by placing it in-between the web toes, brood patch and a warm fold of feathers, where it is cushioned and protected for 9 weeks. The egg remains in that location until it hatches during the coldest months of the Antarctic winter. During this period, the male penguin does not eat and may lose up to half of the body weight; hence, the male penguin must be fat and healthy prior to the breeding season.

In some cases, foster broody hens could be used to incubate eggs from other hens, breeds or species. This is common in local settings where there are no artificial incubation facilities. In Nigeria, for instance, local broody hens are used in incubating guinea fowl, turkey and partridge eggs. In this case, the foster broody hen should be big, in order to cover more eggs during the period, and should have a good brooding and mothering records. Characteristics of such broodiness are that the hen stops laying after a sizeable number of eggs have been laid, remains sitting on the eggs for a longer time and should have enough feathers with a broad broody patch and be able to spread its wings and makes a distinctive clucking sound. In

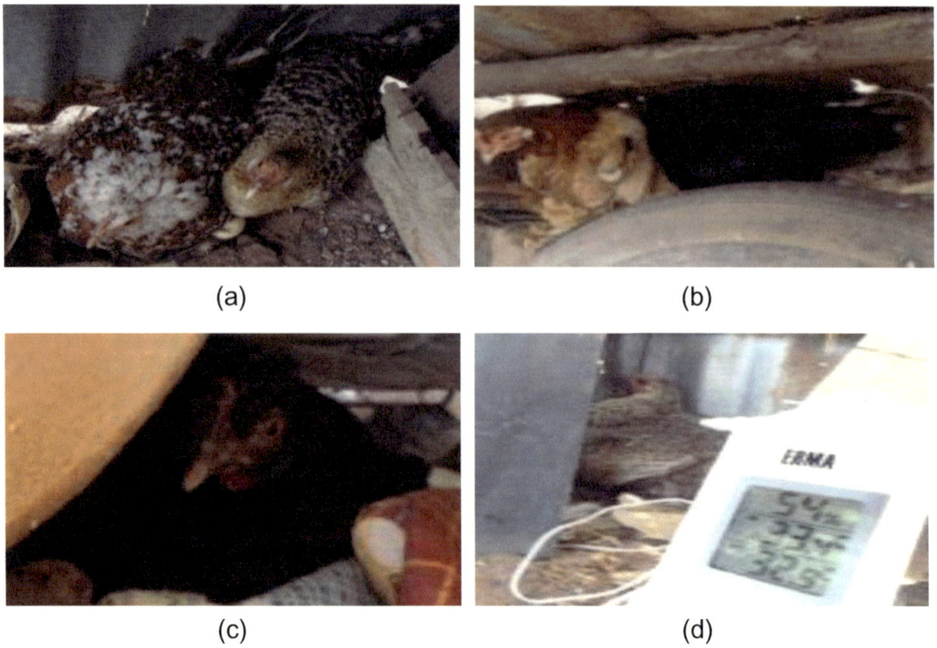

Figure 3.
(a and b) Two-hen mutual broody excerpted from a final year project at NSUK Teaching and Research Farm, Nasarawa State, Nigeria and (c and d) Single-hen broody excerpted from a final year project at NSUK Teaching and Research Farm, Nasarawa State, Nigeria.

some instances, these brooding characteristics may be induced or tested using dummy eggs or even stones. A maximum of 14–16 eggs may be brooded by a foster broody hen, but hatchability often declines with more than 10 eggs, depending on the size of the hen.

Feed and water should be provided in close proximity to the broody hen, in order to keep it in better condition and reduce embryo damage due to the cooling of the eggs, if it has to leave the nest to scavenge for feed and water far away. The hen keeps the eggs at the correct humidity by splashing water on them from her beak. This is a further reason for providing it with easy access to water. In very dry regions, slightly damp soil can be placed under the nesting material to assist the hen in maintaining the correct humidity between 60 and 80%. Fertile eggs from other birds are best added under the brooding hen between one and four days after the start of brooding. In Bangladesh, Sutcliffe [41] reported that local broody hens will even sit on and hatch a second clutch of eggs. However, it often loses considerable weight in the process especially, if sufficient attention was not paid to the provision of food and water. Eggs initially need a very controlled heat input to maintain the optimum temperature, because the embryo is microscopic in size. As the embryo grows in size, it produces more heat than it requires and may even need cooling. Therefore, moisture levels (relative humidity) of 60–80% are important to stop excess moisture loss from the egg contents through the porous eggshell and membranes. According to Sutcliffe [41], there are some factors to consider for a successful natural incubation. These include the following:

- Provision of feed and water close to the broody hen.

- The broody hen should be examined to ensure that it has no external parasites.

- Any eggs stored for incubation should be kept at a temperature between 12 and 14°C at a high humidity of between 75 and 85% and stored for no longer than seven days.

- Extra fertile eggs introduced under the hen from elsewhere should be introduced at dusk.

- The eggs should be tested for fertility after one week by holding them up to a bright light (a candling box works best). If there is a dark shape inside the egg (i.e. the developing embryo), then it is fertile. But if completely clear (translucent), it means the egg is infertile.

- Setting of eggs should be timed so that the hatchlings are two months of age at the onset of major weather changes such as either the rainy or dry season and winter or summer.

- A plentiful natural feed supply over the growing period of the hatchlings should be targeted in order to ensure a better chance for higher survival rate.

Successful avian species instinctively lay and incubate their eggs at a time of the year when the hatchlings will have a better supply of high protein and energy feed sources in the environment. For example, guinea fowl and some species of water fowl (ducks and geese) will only lay and incubate the eggs in the rainy season. However, seasonal changes in weather patterns are also times of greater disease risk. Generally in avian species, a hatchability of 80% of eggs set from natural incubation is normal, but a range of 60–70% is considered satisfactory. The major hindrance in natural incubation is predation especially by hawks, cats, dogs and snakes and environmental hazards particularly uncovered ditches and soil erosion.

1.8.2 Artificial incubation

In artificial incubation, many eggs of similar age (clutch) could be incubated at the same time depending on the capacity of the incubator. The artificial incubator could be homemade or commercial but should typically have heating source, air circulator (fan), temperature regulator (thermostat) as well as water trough and egg trays. Artificial incubator is designed to simulate and mimic the mother hen's role in natural incubation of providing fertile eggs with optimum environmental conditions (temperature, egg turning and humidity) to stimulate embryonic development until hatching [42]. Fertile eggs stored for as long as 7–10 days at room temperature (10–15°C) could be set for incubation. According to ISA [43], eggs should not be incubated the same day it was laid, in order to avoid hatching failure. Woodard et al. [44] stated that on incubation days 0–12, the temperature should be adjusted to 37.5°C, 13–15 days (37.2°C), and on day 16, the temperature should be 37°C and increased to 37.6°C on day 17 when the chicks are expected to emerge. On the other hand, Musa et al. [45] recommended 39.4°C in the tropics, whereas Ferguson [46] gave a range of 37.5–38°C as optimum incubation temperature in poultry production.

It has been shown that environmental temperature is the most important factor in incubation efficiency; therefore a constant incubation temperature of 37.8°C is the thermal homeostasis in avian embryos and gives the best embryo development and hatchability. Thus, incubator temperature should be maintained between 37.2

and 37.7°C [47–49]. Although the acceptable range of incubation temperature varies between 36 and 37°C, Idahor [50] stated that Japanese quail embryo's mortality may be recorded if the temperature drops below 36°C or rises above 40°C for several hours. It was demonstrated that if temperature is at either extreme for several hours, Japanese quail eggs may not hatch. Also, it was observed that overheating was more critical than underheating during the study; hence, running incubator at 40°C for 5 h seriously affected the Japanese quail embryo, whereas running it at 36°C for 5 h only resulted in late hatching. Similarly, Lourens et al. [51] reported significant embryo mortality and lower hatchability in fowl eggs subjected to 38.9°C incubation temperature. Consequently, several researchers have investigated the factors affecting fertility and hatchability in avian species with the sole aim of combating them. For example, it has been shown that age of the breeders, plane of nutrition, mating ratio as well as poor level of management affected fertility in poultry production [52]. Others reported that egg storage conditions, strain of birds, shell quality, season of the year, incubation condition and turning frequency affected hatchability of fertile eggs in poultry birds [53–55]. Recommended optimum incubation temperatures in some avian species are presented in **Table 4** as adapted from Sartell [56].

Avian species	Range of temperature		Typical incubation period (days)
	Celsius (°C)	Fahrenheit (°F)	
Fowl	37.4–37.6	99.3–99.6	21
Guinea Fowl	37.5	99.5	28
Turkey	37.2–37.5	99–99.5	28
Pheasant	37.6–37.8	99.6–100	23–27
Chukar Partridge	37.5	99.5	23
Japanese Quail	37.6–37.8	99.6–100	16–18
Bobwhite Quail	37.5	99.5	22–23
Ducks	37.4–37.6	99.3–99.6	28
Indian Runner Duck	37.5	99.5	28–30
Mallard	37.5	99.5	28–30
Muscovy Duck	37.5	99.5	35–37
Swan	37.5	99.5	30–37
Geese	37.4–37.6	99.3–99.6	28–30
Ostrich	35.8–36.4	96.5–97.5	42
Canada Goose	37.5	99.5	28–30
Egyptian Goose	37.5	99.5	28–30
Emu	35.8–36.1	96.5–97	50–56
Grouse	37.5	99.5	25
Amazons	36.8–37.0	98.3–98.6	24–29
Macaws	36.8–37.0	98.3–98.6	26–28
Love Birds	36.8–37.0	98.3–98.6	22–24
Peafowl	37.5	99.5	26–29
Pigeon	37.5–38.2	99.5–100.5	17
Rheas	35.8–36.4	96.5–97.5	35–40

Sources: [56].

Table 4.
Recommended optimum incubation temperature in some avian species.

1.9 Avian embryogenesis

At fertilization, the avian egg is only fertile, and zygote will only develop outside the body during incubation with appropriate temperature and relative humidity. Fertile eggs begin to develop to embryo when the temperature exceeds physiological zero temperature given as 26–36°C. Below or within this range of temperature, embryonic growth is believed to be halted and at above it (i.e. 36–40.5°C) which is described as the lower limit of optimal development growth is resumed. At above 40.5°C which is the upper lethal temperature, malformation of embryo or embryonic death could occur [57, 58]. According to Boerjan [59], avian embryonic growth could be halted or slowed down and eventually arrested, if the temperature falls below 'physiological zero'. That is the level at which incubation temperature is low enough to keep embryonic cell activity at a greatly reduced rate but reversible level. Essentially, the embryo still has the potential to continue its development again if normal temperature is restored. That is why the term 'arrested development' should be preferred to 'stop development' that is commonly used. As a result, 'physiological zero' should not be restricted to a specific or particular set point temperature, instead to a range of temperature from 12 to 20°C, depending on the milieu of egg handling and storage duration. Hence, the reasons why different set points temperature for 'physiological zero' is defined in different ways depending on the situation being described. The definition of 'physiological zero' was first presented by Edwards [60] as the set point temperature of about 21.0°C, and below this value, there was no embryonic growth. The terms of reference for 'physiological zero' were reviewed by Proudfoot [61] to include a storage temperature range of 11.5–21°C. Fasenko [62] introduced the term 'embryonic diapause' as an alternative to the traditional 'physiological zero' temperature regime. This updated definition recognized that some cellular metabolic processes still continue, but gross morphological changes such as shape and structure are arrested. 'Embryonic diapause' has been described in many vertebrate species such as turtles, marsupials and even mammals such as Roe deer. 'Embryonic diapause' or 'embryonic dormancy' describes a stage at which metabolic activity and cell division are downregulated or arrested and can be regarded as a strategy for coping with temporarily unfavourable environmental conditions. In avian species, embryonic development could be arrested after laying and cooling the eggs down to room temperature of between 22 and 25°C. During this cooling period under optimal conditions, the embryo develops from gastrula stage IX–X as described by Eyal-Giladi and Kochav [63] to stage XII–XIII reported by Gilbert et al. [64].

The definition of physiological zero temperature was restricted specifically to stages XII–XIII of development. If the embryo has developed beyond this stage and primitive streak development has started, reduced temperatures will slow down development and finally result in early mortality of the embryo. This may explain the higher rates of early mortality often recorded, when eggs are kept too long in the nests and when egg cooling is too slow [65, 66]. Therefore, optimum incubation temperature could be given as 37.5–37.6°C but should be reduced to 36.9°C during the last 3 days of the incubation period [44, 67, 68]. Ainsworth et al. [69] stated that in developmental biology experimentations, avian species have been used as models in morphogenesis studies. For example, some staging of Japanese quail embryo development has been attempted but incomplete due to variations in descriptions, staging and incubation processes which were always difficult. It appeared to be a general agreement that at early stages of embryogenesis, there were some developmental differences between fowl embryo [70] and quail embryo [69]. Yet, the basis for these differences has not been established experimentally; hence, Ainsworth [69] constructed a 46-stage series, irrespective of the enhanced ontogeny observed in the Japanese quail in order to make the staging series comparable. At the early stages of development (Stage 4–28), Japanese quail stage series was identical to the Hamburger and Hamilton (HH) stage in fowl chick series as the rate of development of both species was indistinguishable. At the mid stages (Stage 29–35), the descriptions of morphological changes of each stage were still comparable between fowl chick and Japanese quail chick series. At later stages of development (Stage 36–46), the HH stage fowl chick series was no longer comparable to the quail series with regard to incubation periods and morphological descriptions.

It has been reported that biological engineering in avian species has advanced, for example, artificial *in ovo* culture of one-celled zygote of blastoderm stage, made the production of adult bird possible [71]. Similarly, some trials to produce transgenic of chimeric birds have been conducted [72], and the use of avian embryo in teratological test has been anticipated [73]. Nakane and Tsudzuki [74] established that series of normal stages in the development of Japanese quail embryo skeleton composed of 15 stages. In that study, the time of chondrification and calcification of the skeleton were recorded every 24 h from incubation day 3–17 at 37.7°C. It was reported that the knowledge of skeletogenesis stages in Japanese quail embryo will be useful as a normal control not only in experimental embryology, teratology and developmental engineering but also in identifying mutant embryos with skeletal abnormalities. Japanese quail embryo developmental growth was expressed on daily bases by Idahor et al. [75] as culled from NSAP 2018 Proceedings. See **Figure 4a** and **b** as given by Hamburger and Hamilton [70] and Idahor et al. [75].

1.10 Hatchability

The fully grown embryo uses the egg tooth to carefully window the eggshell and emerges as day-old hatchling. At hatch, some avian species could walk and scavenge on their own; hence, they are referred to as precocial birds. Examples are ostriches, geese, turkeys, ducks, fowls, pheasants and partridges. Whereas, others such as golden eagles, doves, pigeons, starlings, robins, wrens and hummingbirds cannot walk or fend for themselves; hence, they are regarded to as altricial birds; thus, they depend on the dam and/or sire for survival. Some factors such as age of breeders, mating system, eggs set and storage conditions, incubation temperature, ventilation, relative humidity and egg turning angle may affect hatchability [47, 51, 76]. Since incubation temperature has been described as the most critical environmental concern during hatchery operations.

Table 5 shows the most appropriate time to transfer incubated eggs to the hatcher compartment in the incubator, in order to achieve the recommended 60–70% hatchability. Hatchability could be determined using:

$$\text{Hatchability} = \frac{\text{Total number of hatchlings}}{\text{Total number of fertile eggs}} \times \frac{100}{1} \quad (1)$$

1.11 Chick sex identification

In avian species, sex identification is critical because in egg-laying operations, the male hatchlings are irrelevant, thus should be discarded. At hatch, all the identified males are supposed to be destroyed but could be reared as poussins for meat. However, the time and resources required to rear them are apparently a waste except in a free-range system. Nevertheless, at hatch, the sex of the hatchling may be identified in some species, whereas it may be impossible until several days of post-hatch in some other species. With hi-tech facilities, hatchling sexing could be done in various ways but sex-linked genes where plumage pattern or wing

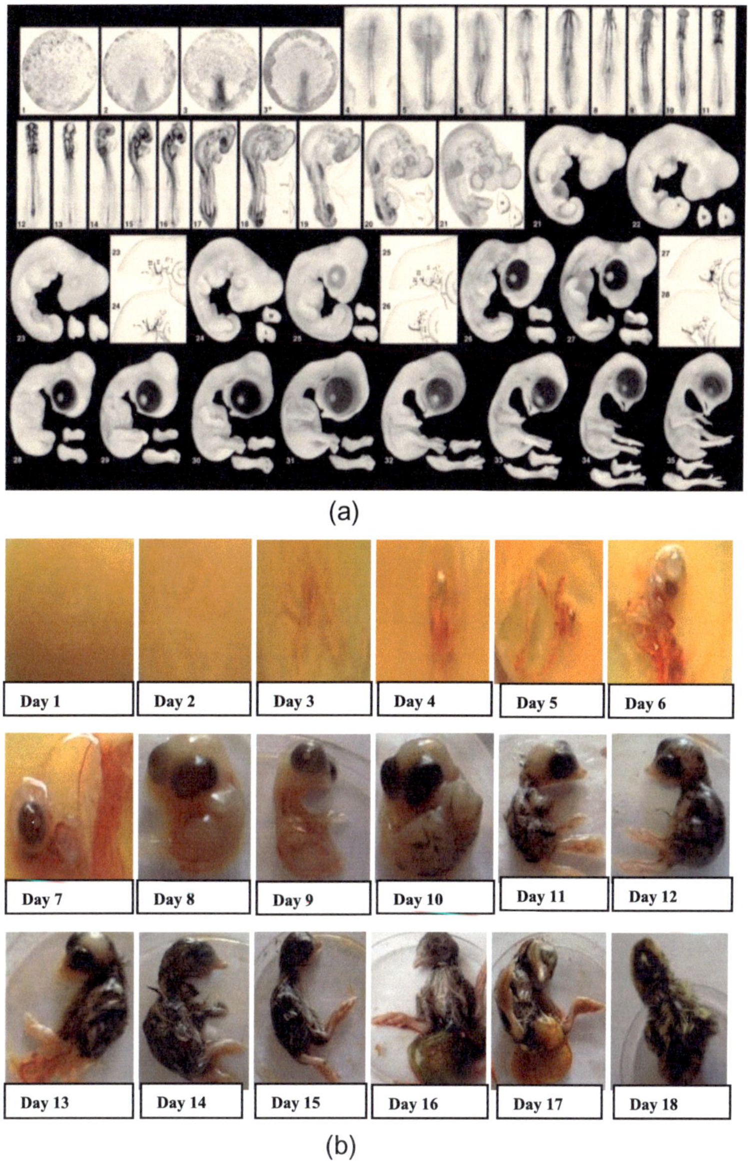

Figure 4.
(a) Stages of chick embryonic development adapted from Hamburger and Hamilton [70] and (b) Japanese quail embryo developmental stages culled from NSAP 2018 Proc. as given by Idahor et al. [75].

feathering is used to identify hatchling sexes at a day old is commoner. Even with the hi-tech facilities, hatchling sex identification in Japanese quail is impossible until about 3 weeks of age. According to Campbell et al. [23], some other sexing techniques in avian species include cloacal examination to view the rudimentary papillae and proctoscope to locate the testes. The sexed avian species should be reared as hatchlings, growers, finishers and breeders to meet the aim of the farmer.

1.12 Conclusion

Most avian species have brilliant senses and flight capability though few flightless species exist. They possess wings, beaks and lightweight hollow skeleton. Some of the avian species have been domesticated while others are still in the wild. In any

Common name	Incubation period (days)	Incubation conditions		Hatcher conditions		
		Temp (°F)	R/H (%)	Transfer day	Temp (°F)	R/H (%)
Canary	13–14	100.5	56–58	11	99	66–74
Chicken	21	99.5	58	18	98.5	66–75
Cockatiel	18–20	99.5	58–62	15–18	99	66–74
Cockatoo	22–30	99.5	58–62	20–27	99	66–74
Conure (sun)	28	99.5	58–62	25	99	66–74
Conure (various)	21–30	99.5	58–62	18–27	99	66–74
Dove	14	99.5	58	12	98.5	66–75
Duck	28	99.5	58–62	25	98.5	66–75
Muscovy duck	35–37	99.5	58–62	31–33	98.5	66–75
Finch	14	99.5	58–62	12	99	66–74
Domestic goose	30	99.5	62	27	98.5	66–75
Geese (various)	22–30	99.5	62	20–27	98.5	66–75
Grouse	24–25	99.5	54–58	22	99	66–74
Guinea	28	99.5	54–58	22	99	66–74
Lovebird	22–25	99.5	58–62	20–22	99	66–74
Macaw	26–28	99.5	58–62	23–25	99	66–74
Mynah	14	100.5	56–58	12	99	66–74
Parakeet	18–26	99.5	58–62	15–23	99	66–74
Budgerigar	18	99.5	58–62	15	99	66–74
Parrot (various)	18–28	99.5	58–62	15–25	99	66–74
Parrot (African grey)	28	99.5	58–62	25	99	66–74
Chukar partridge	23–24	99.5	62	20	99	66–74
Peafowl	28–29	99.5	58–62	25–26	98.5	66–75
Ptarmigan	21–23	99.5	58–62	18–20	99	66–74
Raven	20–21	99.5	58–62	17–18	99	66–74
Ring-neck pheasant	24–24	99.5	58–62	21	99	66–74
Pheasant	22–28	99.5	58–62	20–25	99	66–74
Pigeon	17–19	100.5	58	14	99	66–74
Bobwhite quail	23	99.5	54–58	21	99	66–74
Japanese quail	17–18	99.5	58–62	15	99	66–74
Swan	33–37	99.5	58–62	30–33	99	66–74
Turkey	28	99.5	54–58	25	98.5	66–75
Emu	49–50	97.5	32–40	47	97.5	69
Ostrich	42	97.5	32–40	39	97.5	69
Rhea	36–42	97.5	50	34–37	97.5	69

Source: [77].

Table 5.
Incubation period and when to transfer to hatcher in some avian species.

case, both wild and domesticated are relevant to human existence as they complement the ecosystem for enhanced ecological sustainability. They are essentially classified as egg-type, meat-type or dual-purpose type. They have varied chromosome numbers, and the sex is determined by the female because it is heterogametic. All the males possess paired testes that are retained within the abdominal region, and some of them have intromittent organ while few others have penis. The female has a single ovary capable of producing so many eggs in a breeding season. While many reproduce all year round, a few are seasonal breeders. Egg production is not dependent on mating but infertile eggs will be laid purely for human consumption, whereas mating is required for fertile eggs that must be incubated for procreation. The egg is very rich in nutrients, and its biological value is approximately 100%. Avian species can live for several years except a few that can live for less than 5 years. Consequently, knowledge of avian reproduction is sacrosanct to protect avian species all over the world through intensive domestication process, deliberate breeding strategies among other measures to enhance their sustainability.

Author details

Kingsley Omogiade Idahor
Department of Animal Science, Nasarawa State University, Lafia, Nigeria

*Address all correspondence to: koidahor@nsuk.edu.ng

References

[1] Lederer R, Burr C. Latein für Vogelbeobachter: Über 3000 ornithologische Begriffe erklärt und erforscht, aus dem Englischen übersetzt von Susanne Kuhlmannn-Krieg. Köln: Verlag DuMont; 2014. ISBN 978-3-8321-9491-8

[2] Xu X, Zhou Z, Wang X, Kuang X, Zhang F, Du X. Four-winged dinosaurs from China. Nature. 2003;**421**(6921): 335-340. DOI: 10.1038/nature01342

[3] National Geographic. Field Guide to Birds of North America. 7th ed. National Geographic; 2017. ISBN: 9781426218354.

[4] Jones DN, Dekker R, Roselaar WRJ, Cees S. The Megapodes. Bird Families of the World 3. Oxford: Oxford University Press; 1995. ISBN 0-19-854651-3

[5] Osinowo OA. Introduction to Animal Reproduction. Abeokuta: Sophie Academic Services Ltd.; 2006. p. 91

[6] Oluyemi JA, Roberts FA. Poultry Production in Warm Wet Climates. 2nd ed. Ibadan, Nigeria: Spectrum Books Ltd; 2000. p. 244

[7] Johnson AL. Reproduction in the male. In: Sturkie PD, editor. Avian Physiology. New York, NY: Springer; 1986

[8] Brennan PLR, Prum RO. The erection mechanism of the ratite penis. Journal of Zoology. 2011. DOI: 10.1111/j.1469-7998.2011.00858.x. Available from: https://prumlab.yale.edu/sites/default/files/brennan_prum_2011.pdf

[9] Montgomerie R. Sexual conflict and the intromittent organs of male birds. In: Leonard JL, Córdoba-Aguilar A, editors. The Evolution of Primary Sexual Characters in Animals. New York: Oxford University Press; 2010. pp. 453-470. ISBN 9780199886753

[10] Birkhead PLR, Prum RO, McCracken KG, Sorenson MD, Wilson RE, Birkhead TR. Coevolution of male and female genital morphology in waterfowl. PLOS ONE. 2007;**2**(5): e418. DOI: 10.1371/journal.pone.0000418

[11] Marcus A. Ostrich penis clears up evolutionary mystery. Nature. 2011;**2011**. DOI: 10.1038/nature.2011.9600

[12] Johnson L, Varner DD, Roberts ME, Smith TL, Keillor GE, Scrutchfield WL. Efficiency of spermatogenesis: A comparative approach. Animal Reproduction Science. 2000;**60-61**: 471-480

[13] Aire TA. Spermatogenesis and testicular cycles. In: Jamieson BGM, editor. Reproductive Biology and Phylogeny of Birds, Part A, Phylogeny, Morphology, Hormones and Fertilization. Enfield, New Hampshire, USA, and Plymouth, UK: Science Publishers, Inc.; 2007

[14] Lin M, Jones RC. Spermiogenesis and spermiation in the Japanese quail (*Coturnix coturnix japonica*). Journal of Anatomy. 1993;**183**:525-535

[15] Tiba T, Yoshida K, Miyake M, Tsuchiya K, Kita I, Tsubota YT. Regularities and Irregularities in the structure of the seminiferous epithelium in the domestic fowl (*Gallus domesticus*) — I. Suggestion of the presence of the seminiferous epithelial cycle. Anatomy, Histology and Embryology. 1993;**22**: 241-253

[16] Sprando IL, Russell LD. Spermiogenesis in the red-eared turtle (*Pseudomys scripta*) and the domestic fowl (*Gallus domesticus*): A study of cytoplasmic events including cell volume changes and cytoplasmic elimination. Journal of Morphology. 1988;**198**:95-118

[17] Adeoye GO, Oleforuh-Okoleh VU, Chukwuemeka UM. Influence of breed type and age on spermatological traits of Nigerian local chickens. Agro-Science. 2017;**161**:11-16. DOI: 10.4314/as.v16i1.3

[18] Almahdi AB, Ondho YS, Sutopo. Comparative studies of semen quality on different breed of chicken in poultry breeding center Temanggung-Central Java. International Refereed Journal of Engineering and Science. 2014;**3**:94-103

[19] Fitzpatrick FL. Bilateral ovaries in raptorial birds, with notes on kidney structure. Proceedings of the Iowa Academy of Science. 1931;**38**(1):245-248

[20] Rodler D, Stein K, Korbel R. Observations on the right ovary of birds of prey: A histological and immunohistochemical study. Anatomia Histologia Embryologia. 2014;**44**:168-177

[21] Suyatma NE, Hermanianto J. Egg Characteristics (4th Revision). Institut Pertanian: Bogor Agricultural University; 2010

[22] Pollock CG, Orosz SE. Avian reproductive anatomy, physiology and endocrinology. The Veterinary Clinics of North America. Exotic Animal Practice. 2002;**5**(3):441-474

[23] Campbell JR, Kenealy MD, Campbell KL. Animal Sciences: The Biology, Care and Production of Domestic Animals. New Yolk: McGraw-Hill Inc.; 2003. p. 510

[24] Brillard JP, Antoine H. Storage of sperm in the uterovaginal junction and its incidence on the numbers of spermatozoa present in the perivitelline layer of hens' eggs. British Poultry Science. 1990;**31**:635-644

[25] Tingari MD, Lake PE. Ultrastructural studies on the uterovaginal sperm-host gland of the domestic hen, *Gallus domesticus*. Journal of Reproduction and Fertility. 1973;**34**:423-431

[26] van Krey HP, Ogasawara FX, Pangborn J. Light and electron microscope studies of possible sperm gland emptying mechanisms. Poultry Science. 1967;**46**:69-78

[27] Matsuzaki M, Hiyama G, Mizushima S, Shiba K, Inaba K, Sasanami T. Specific mechanism of sperm storage in avian oviducts. In: Sawada H, Inoue N, Iwano M, editors. Sexual Reproduction in Animals and Plants. Tokyo: Springer; 2014

[28] Ito T, Yoshizaki N, Tokumoto T, et al. Progesterone is a sperm releasing factor from the sperm storage tubules in birds. Endocrinology. 2011;**152**:3952-3962

[29] Holt WV. Does apoptosis hold the key to long-term sperm storage mechanisms *in vivo*? Molecular Reproduction and Development. 2011;**78**:464-465

[30] Holt WV, Lloyd RE. Sperm storage in the vertebrate female reproductive tract: How does it work so well? Theriogenology. 2010;**73**:713-722

[31] Baer B, Eubel H, Taylor NL, O'Toole N, Millar AH. Insights into female sperm storage from the spermathecal fluid proteome of *Apis mellifera*. Genome Biology. 2009;**10**(6): R67

[32] Baer B, Armitage SAO, Boomsma JJ. Sperm storage induces an immunity cost in ants. Nature. 2006;**441**(7095):872-875

[33] Bakst MR. Role of the oviduct in maintaining sustained fertility in hen. Journal of Animal Science. 2011;**89**:1323-1329. DOI: 10.2527/jas.2010-3663

[34] Bakst MR, Wishart G, Brullard JP. Oviductal sperm selection, transport,

and storage in poultry. Poultry Science Reviews. 1994;**5**:117-143

[35] Brillard JP. Sperm storage and transport following natural mating and artificial insemination. Poultry Science. 1993;**71**:923-928

[36] Adeyina AO, Akanbi AD, Ibiwoye KO, Fatai AO, Adeyina OA. Effects of mating period on egg fertility and egg characteristic of Japanese quails. In: Proc. 5th Intl Poult Summit. Nigeria: UNILORIN; 2015. pp. 222-225

[37] Bakst MR, Donoghue AM, Yoho DE, Moyle JR, Whipple SM, Camp MJ, et al. Comparisons of sperm storage tubule distribution and number in 4 strains of mature broiler breeders and in turkey hens before and after the onset of photostimulation. Poultry Science. 2010;**89**(5):986-992

[38] Birkhead TR, Moller AP. Numbers and size of sperm storage tubules and the duration of sperm storage in birds: A comparative study. Biological journal of the Linnean Society. 1992;**45**:363-372

[39] Birkhead TR, Fletcher F. Sperm storage and release of sperm from the sperm storage tubules in the Japanese quail (*Coturnix japonica*). Comments. 1994;**136**:101-105

[40] Brian B. How to incubate chicken eggs. 2015. Available from: http://modernfarmer.com

[41] Sutcliffe JH. Natural and artificial incubation. 1909. Available from: http://www.archive.org/details/cu31924003170325

[42] French NA. Modelling incubation temperature: The effect of incubation design on embryonic development and egg size. Poultry Science. 1997;**76**:124-133

[43] ISA. ISA Brown Commercial Stock. A Hendrix Genetics Co; 2016. Available from: http://www.isapoultry.com/en/Products/ISA/ISA%20Brown.aspx

[44] Woodard AE, Abplanalp H, Wilson WO, Wohra P. Japanese quail husbandry in the laboratory. Department of Avian Sciences: University of California; 1973. p. 22

[45] Musa U, Haruna ES, Lombin LH. Quail Production in the Tropics. Vom: NVRI Press; 2007. p. 158

[46] Ferguson MWJ. Temperature dependent sex determination in reptiles and manipulation of poultry sex by incubation temperature. In: Proceedings 9th European Poultry Conference. Glasgow, UK; 1994. pp. 380-382

[47] Hill D. Chick length uniformity profiles as a field measurement of chick quality. Avian and Poultry Biology Reviews. 2001;**12**:188

[48] Lourens A, van den Brand H, Hectkamp MJW, Meijerhof R, Kemp B. Effects of eggshell temperature and oxygen concentration on embryo growth and metabolism during incubation. Poultry Science. 2007;**86**:2194-2199

[49] Wilson RH. Effect of egg size on hatchability, chick size and post-hatching growth. In: Hocking PM, editor. Avian Incubation. Wallingford, Oxfordshire, UK: CAB International; 1991. pp. 279-283

[50] Idahor KO. Optimum incubation temperature determination and alteration to enhance sex reversal, hatchability and post-hatch performance of Japanese quails (*Coturnix coturnix japonica*). In: A PhD thesis submitted to the Faculty of Agriculture and Forestry, Department of Animal Science. Nigeria: University of Ibadan; 2021. p. 284

[51] Lourens A, van den Brand H, Meijerhof R, Kemp B. Effects of eggshell temperature during incubation on embryo development, hatchability and

post-hatch development. Poultry Science. 2005;**84**:914-920

[52] Ishola OO, Ogundipe GAT. Causes and financial implications of reduced hatchability of chicken eggs at a commercial hatchery in Abeokuta, Nigeria. Nigeria Veterinary Journal. 1998;**19**:12-20

[53] Benneth CD. The influence of shell thickness on hatchability in commercial broiler breeder flocks. Journal of Applied Poultry Research. 1992;**1**:61-65

[54] Jayarajan S. Seasonal variation in fertility and hatchability of chicken eggs. Indian Journal of Poultry Science. 1992;**27**:36-39

[55] Obioha FC, Okorie AU, Akpa MO. The effect of egg treatment, storage and duration on hatchability of broiler eggs. Archiv Fur Geflugelkunde. 1986;**50**: 213-218

[56] Sartell J. Incubation station: A reference guide to incubation terms, temperatures, times and humidity levels. Community Chickens Newsletter. 2018. Available from: https://www.communitychickens.com/category/coops/; https://www.beautyofbirds.com/eggincubationtemperature.html

[57] Conway JC, Thomas EM. Effects of ambient temperature on avian incubation behaviour. Behavioural Ecology. 2000;**11**:178-188

[58] Webb DR. Thermal tolerance of avian embryos: A review. The Condor (The Cooper Ornithological Society). 1987;**89**:874-898

[59] Boerjan M. A Practical interpretation of 'physiological zero' in hatchery management. 2016. p. 41

[60] Edwards CL. The physiological zero and the index of development for the egg of the domestic fowl, Gallus domesticus. American Journal of Physiology. 1902;**6**:351-397

[61] Proudfoot FG. The handling and storage of hatching eggs. In: Carter TC, Freeman BM, editors. The Fertility and Hatchability of the Hen's Egg. Edinburgh: Oliver and Boyd; 1969. pp. 127-141

[62] Fasenko GM. Egg storage and the embryo. Poultry Science. 2007;**86**: 1020-1024

[63] Eyal-Giladi E, Kochav S. From cleavage to primitive streak formation: A complementary normal table and a new look at the first stages of the development of the chick. I. General morphology. Developmental Biology. 1976;**49**:321-337

[64] Gilbert DC, Norman AR, Nicholl J, Dearnaley DP, Horwich A, Huddart RA. Treating stage I nonseminomatous germ cell tumours with a single cycle of chemotherapy. Journal Compilation International. 2006;**98**:67-69. DOI: 10.1111/j.1464-410X.2006.06188.x

[65] Fasenko GM, Robinson FE. Profiling egg storage: The effects on egg weight loss, egg characteristics and hatchability. Poultry Science. 1999;**78**(Suppl. 1):9. Abstract

[66] Fasenko GM, Robinson FE, Armstrong JG, Church JS, Hardin RT, Petitte JN. Variability in preincubation embryo development in domestic fowl. 1. Effects of nest holding time and method of egg storage. Poultry Science. 1991;**70**:1876-1881

[67] Harb SK, Habbib YA, Kassem AM, El-Raies A. Energy consumption for poultry egg incubator to suit small farmer. Egyptian Journal of Agricultural Research. 2010;**88**(1):193-210

[68] Scott HS, Steven RB. Egg viability as a constraint on hatching synchrony at high ambient temperatures. Journal of Animal Ecology. 1999;**68**:951-962

[69] Ainsworth SJ, Stanley RL, Evans DJR. Developmental stages of the

Japanese quail. Journal of Anatomy. 2010;**216**:3-15. DOI: 10.1111/j.1469-7580. 2009.01173.x

[70] Hamburger V, Hamilton HL. A series of normal stages of development of the chick embryo. Journal of Morphology. 1951;**88**:4962

[71] Ono T, Murakami T, Mochii M, et al. A complete culture system for avian transgenesis, supporting quail embryos from the single-cell stage to hatching. Developmental Biology. 1994a;**161**: 126-130

[72] Ono T, Muto S, Mizutani M, et al. Production of quail chimera by transfer of early blastodermal cells and its use for transgenesis. Japan Poultry Science. 1994b;**31**:119-129

[73] Hashizume R, Noda A, Itoh M, Yamamoto Y, Masui S, Oka M. A method for detecting malformations in chicken embryos. Japan Poultry Science. 1993;**30**:298-305

[74] Nakane Y, Tsudzuki M. Development of the skeleton in Japanese quail embryos. Development, Growth & Differentiation. The Japanese Society of Developmental Biologists. 1999;**41**(5): 523-534

[75] Idahor KO, Sokunbi OA, Osaiyuwu OH, Odu O, Nwosuh CI, Sati NM, et al. Embryo developmental stages in Japanese quail (*Coturnix coturnix japonica*) eggs incubated at 38°C. In: Proc. 43rd Ann Conf NSAP, FUT Owerri. 2018. pp. 1433-1436

[76] Permsak S. Effect of water spraying and eggs turning angle to efficiency of duck hatchability. In: Proc 34th Kasetsart University Ann Conf Bangkok (Thailand). 1996. pp. 22-26

[77] Archer GS, Cartwright AL. Incubating and Hatching Eggs. Texas A&M AgriLife Extension Service, AgriLifeBookstore.org. EPS-001 7/13; 2018

Section 5

Reproduction in Aquatic Animals

Chapter 7

Stimulatory Effects of Androgens on Eel Primary Ovarian Development - from Phenotypes to Genotypes

Yung-Sen Huang and Chung-Yen Lin

Abstract

Androgens stimulate primary ovarian development in Vertebrate. Japanese eels underwent operation to sample the pre- and post-treated ovarian tissues from the same individual. Ovarian phenotypic or genotypic data were mined in a pair. A correlation between the initial ovarian status (determined by kernel density estimation (KDE), presented as a probability density of oocyte size) and the consequence of androgen (17MT) treatment (change in ovary) has been showed. The initial ovarian status appeared to be important to influence ovarian androgenic sensitivity. The initial ovary was important to the outcomes of androgen treatments, and ePAV (expression presence-absence variation) is existing in Japanese eel by analyze DEGs; core, unique, or accessory genes were identified, the sensitivities of initial ovaries were correlated with their gene expression profiles. We speculated the importance of genetic differential expression on the variations of phenotypes by 17MT, and transcriptomic approach seems to allow extracting multiple layers of genomic data.

Keywords: androgen, *Anguilla japonica*, ePAV, kernel density estimation, primary ovary, transcriptome

1. Introduction

Anguilla eels, *Anguilla spp.*, are important economic species, but eel seeds (glass eels) for aquaculture are obtained only from the wild. Moreover, the resource of glass eels has been drastically reduced since the 1970s. The artificial propagation of eel seed is demanded. The complex life cycle of Anguilla eels, undergo several metamorphoses and migrate between freshwater and ocean environments, leads an obstacle to the artificial propagation on a commercial scale, although progress in the artificial maturation of eel broodstocks and larvae rearing has been achieved for Japanese eels [1–3]. Indeed, to obtain eel seeds in captivity, an important factor is the selection of broodstocks for artificially maturation [2]. It has been demonstrated that wild silver eels (undergoing pubertal metamorphosis) are the best candidates for artificial maturation [4]. However, natural sources of wild female silver eels are diminishing.

There are different responsive abilities of different eels on piscine pituitary extracts (PPE) in artificial maturation manipulation. This phenomenon has been commented: "the maturity status of parental females at the beginning of artificial maturation trials is also an important factor affecting the success of maturation experiments" [2] as well as suggested that silver eels with more developed oocytes (mean of oocyte diameter is 0.262 mm) required a shorter time to produce mature eggs than did cultivated eels (mean of that is 0.177 mm) [4], the maturity of morphological changes (silvering, as a pubertal metamorphosis in eels) might have a close correlation with ovarian development (oocyte growth). The physiological status of the eels just before PPE injection has been postulated to be a reason for this difference [3]. In addition, in most circumstances during artificial eel maturation, unsynchronized development stages of ovary are a problem, androgens have been demonstrated to synchronize oocyte development in the eel ([5] references therein). While, in practice, the stimulatory effect of androgens (or PPE) on eel ovarian development was often not uniform or not conserved among different individuals, and the stimulatory effects might depend on the initial status of ovarian development [3, 4]. Artificial maturation in eels appears to be paralogue to 'personalized medicine' in human diseases. Identifying a set of gene biomarkers to predict or estimate the outcomes of eel artificial maturation is interesting and is demanded.

2. From phenotypes

Indeed, the developmental stages of oocytes in Japanese eels were classified by the mean of oocyte diameter as follows: oil droplet stage for follicles <200 μm; early vitellogenic stage for follicles 200–400 μm; mid-vitellogenic stage for follicles 400–600 μm; late vitellogenic stage for follicles 600–800 μm; and migratory nucleus stage for follicles 800–1000 μm [4, 6].

Androgens stimulate the development of the preantral follicle in mammals [7]. In fish, androgens participate in the promotion of primary and/or early secondary ovarian follicle growth (reviewed by [8]), the primary and/or early secondary ovarian follicle stages in the fish is analogous to the preantral follicle in mammals. The stimulatory effect of androgens has been reported in coho salmon (*Oncorhynchus kisutch*) *in vivo* [8]. Androgens also significantly promote primary growth (PG) and the pre-vitellogenic (PV) transition in eels *in vivo* (*Anguilla japonica*: [9]; *Anguilla australis*: [10]). Furthermore, androgen treatment is recommended prior to or during eel artificial maturation (*Anguilla japonica*: [5] references therein; *Anguilla australis*: [10]; *Anguilla anguilla*: [11]).

Indeed, artificial maturation is often unsuccessful when the eel broodstocks come from aquafarms or have a larger body size [12]. Whether, the hypotheses that the outcomes of artificial eel maturation are based on "the maturity status of parental females at the beginning of artificial maturation trials" [4] and "the physiological status of the eels just before hormone" [3] are accepted. Therefore, identifying sex and evaluating gonadal maturity are necessary.

A method to track ovarian development before- and after-androgen (17α-methyltestosterone, 17MT) treatment in the same eel has been developed [5]. It must be emphasized that the experimental data came from the same individual; this traceability was the advantage, but the bias caused by a single sample could not be avoided. The pretreated ovary was used as a morphological or molecular baseline to compare the posttreatment ovary. Since the experimental eels and PPE were both from wild populations and subject to important individual variation, both of them are not consisted. This means the results based on two important variations if the different eels as well as PPE were employed simultaneously. To increase sample size can reduce the variation (to approach normal distribution), it can be speculated that the normal distribution is composed of many eels, the number of eels will be numerous, when one eel was selected, since time cannot be reversal, the normal distribution will collapse into one sample, then the destinations or outcomes of the selected one might be an another distribution (not always a normal distribution), but, apparently, there should be a closer correlation between the inputs and the outcomes. The data followed this protocol can be accumulated then be clustered, different patterns or clusters related with certain biological outcomes may be distinguished.

Eels were surgically operated to identify sex and to sample ovarian tissues, the operated eels recovered and the sutures were shed after 4–5 weeks [5, 13]. The manipulation did not suspend ovarian development under the subsequent artificial maturation regimen.

Primary ovarian sensitivity to androgens (17MT) is well documented; this characteristic was used to study the correlation between basal ovarian status and the outcome in Japanese eel. The lard grease- and Vaseline-based sustained-release mixtures were used *in vivo* to stimulate ovarian development. The recovered eels were treated with lard grease-Vaseline +17MT (LVMT) for four weeks. Classically, the ovarian status was estimated by the measurement of the oocyte size in the sampled ovarian tissue, then the distribution of oocyte size was plotted in histogram, histogram is a description method, the mean (m) and standard error (SE) of oocyte size were calculated (e.g. [4]). Kernel density estimation (KDE), a nonparametric way to estimate the probability density function of a random variable, was used to evaluate the ovarian status; one of the advantages of KDE is, compared with histogram method, the tiny differential varieties on the ovarian status can be percept; indeed, the measurement of oocyte diameter under microscope is a daunt task, many conditions can aberrate the results, the oocytes might be deformed by the manipulation of sampling process and tissue fixed. The oocyte diameter was derived from digital data (virtual measurements), we could not deny that the fidelity of 'real' oocyte diameter might be lost and erratic, but this shortcoming could be ameliorated by KDE, Kernel smoothing refers to a general methodology for the recovery of underlying structures in datasets [14].

In the pretreatment eels, there was no significant ($p > 0.05$) difference among them on mean of oocyte diameter, the developmental stage of our experimental eels was the oil droplet stage (primary oocyte). After 4 weeks of treatment (posttreatment), The mean diameter of oocyte significantly ($p < 0.05$) increased in the same

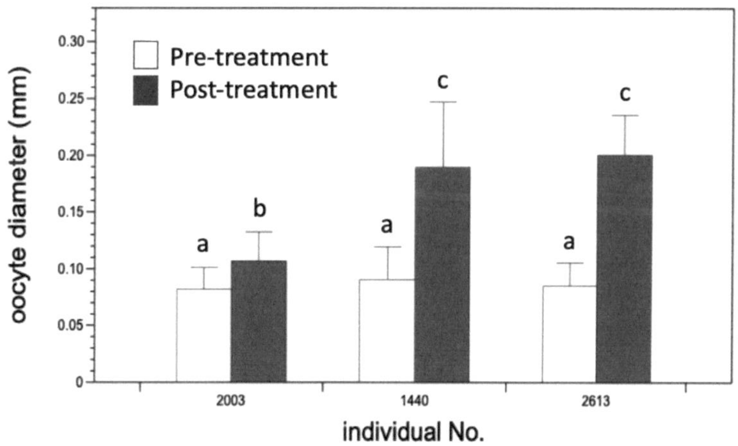

Figure 1.
Mean oocyte diameters of different individuals treated with LVMT. Open or solid columns represent pretreatment or posttreatment, respectively. Different letters represent significant differences (p < 0.05) among different data.

individual, and the mean of fold change was x1.25. Based on KDE and using 0.1 mm as the arbitrary boundary, the percentage of the oocyte diameter over 0.1 mm increased from 44–70% [5].

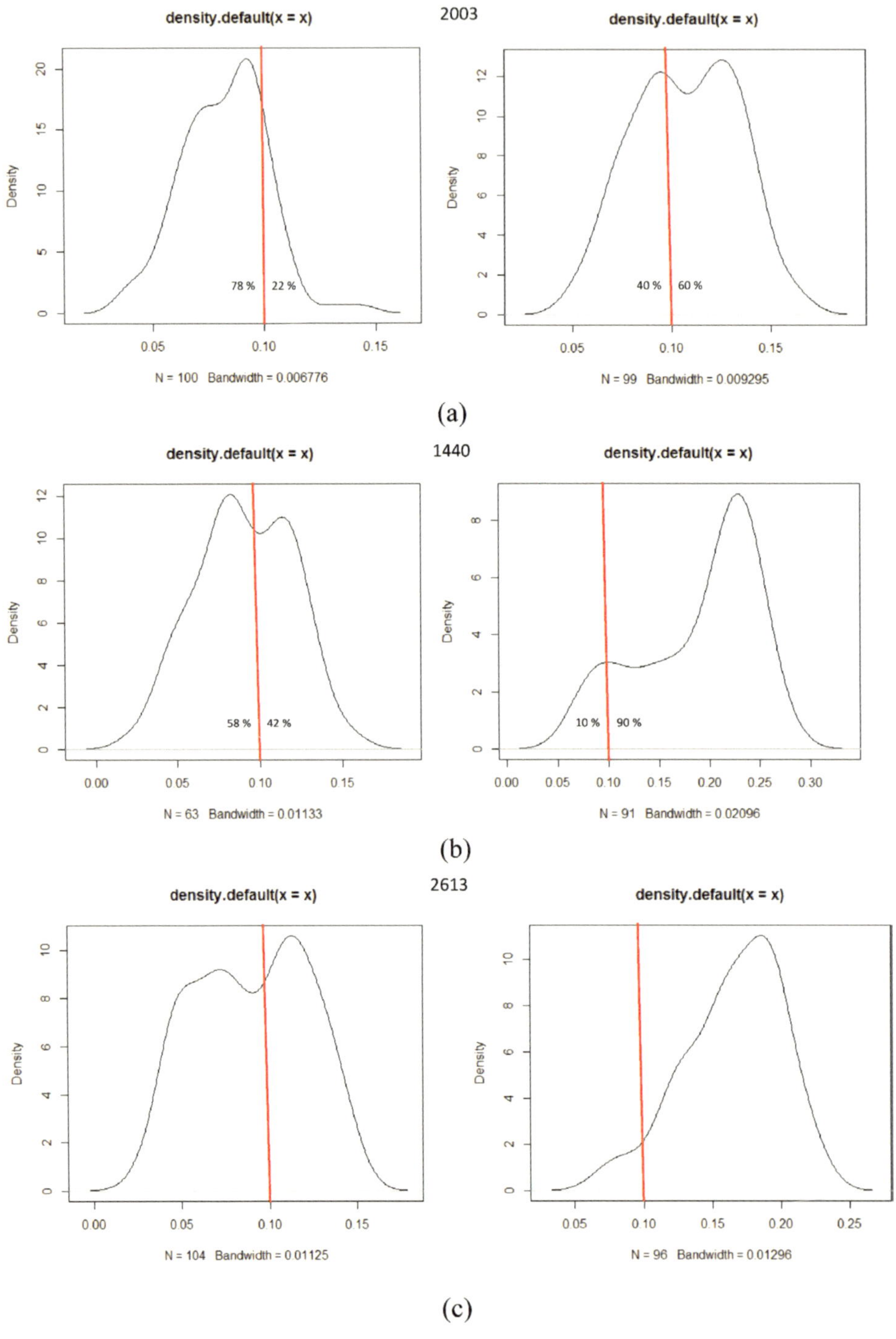

Figure 2.
Estimation of ovarian status by kernel density estimation (KDE). The x-axis and y-axis represent the oocyte diameter (mm) and probability density, respectively. 0.1 mm was chosen as an arbitrary boundary. The distribution of oocyte sizes from the eel no 2003, no 1440 , and 2613 are presented in (a), (b), and (c), respectively.

Three eels, with different fold change on oocyte diameter x1.30, x2.08, and x2.35 tagged with no 2003, no 1440, and no 2613, respectively, were chosen to present the different results of the17MT treatment, although the growth of all oocytes was significantly ($p < 0.05$) stimulated by LVMT, as revealed by pre- vs. posttreatment comparisons (**Figure 1**). Moreover, analyzed by KDE, and 0.1 mm was used as an arbitrary boundary to evaluate ovarian status, the percentage of the oocytes over 0.1 mm increased from 22–60% in eel no 2003, from 42–90% in eel no 1440, and from 40–89% in eel no 2613. It seemed that oocyte growth (development) was more synchronized in eel no 2613 than in eel no 1440 because the bandwidth in eel no 2613 (bandwidth = 0.009722) was smaller than that in eel no 1440 (bandwidth = 0.02096), which indicates a more dispersed trend in ovarian development in eel no 1440; these data corresponded to the standard deviations (SDs) of 0.0574 and 0.0352 for eel no 1440 and no 2613, respectively (**Figure 2**) [5].

KDE may be a way to mine the "maturity status" of the ovary. From plotted figure, if we chose 0.1 mm as the boundary, we can speculate that if the main peak was located on or right-hand to the boundary, better oocyte development results could be detected under the subsequent treatment condition with 17MT.

More than two peaks can identified in certain plotted figures, a shoulder instead of a smooth bell curve was also observed, hidden peak was likely present [5]. Our observed data implied a wave-like on the development of Japanese eel oocytes [5]; coincidently, the Japanese eel might spawn more than once during a spawning season [15]. Moreover, there is also an argument that if the Japanese eel is a not semelparous (monocyclic) fish [16].

The pretreated ovary was used as a morphological or molecular baseline to compare the posttreatment ovary in the same individual. The comparison was based on the pattern from a probability density function (KDE); the patterns represented ovarian status. Our results showed that a different ovarian status seemed to lead to a different artificial maturation results [5].

3. To genotypes

In Japanese eel, the correlations between the initial ovarian status, represented by the probability distributions of oocyte diameter, and the effects of 17MT treatment is shown [5]. We continued to investigate the relationships between the initial ovarian gene expression profiles and the stimulatory effects of androgens on oocyte development in eels. As the activation and development of ovaries are sophisticated (reviewed by [17, 18]), myriad factors and ample genes interact with each other to control folliculogenesis [19]. Following this concept, we studied this phenomenon based on the ovarian expression profiles of the initial ovaries using transcriptomic tools, high-throughput methods for the large-scale analysis of ovarian gene functions can be used to approach this question.

Gene set analysis (GSA) has the advantage of incorporating existing biological knowledge into the expression analysis (reviewed by [20], references therein). Gene Ontology (GO) terms can be used to define gene sets, and GO analysis is a standard way to define the relationships among expressed genes, thus enabling identification through the use of GSA. Both GO and COG (Cluster of Orthologous Genes) provide specific information about genes or gene products.

In teleosts, there is a lack of transcriptome analysis on early folliculogenesis (reviewed by [21]). In coho salmon (*Oncorhynchus kisutch*), 11-ketotestosterone promoted the growth of primary ovarian follicles and altered the ovarian transcriptome dramatically [8]. In Japanese eel, ovarian transcriptomic information from eels treated with salmon pituitary extracts has been reported [22].

The number of total trinity transcripts was 541,020, and 452,778 transcripts were derived from raw assembled transcripts but clustered with 95% similarity. The number of total trinity 'genes' was 336,973, while a total of 25,340 unigenes were distributed into the 3 categories with 37.56% in biological processes, 49.15% in molecular functions, and 13.28% in cellular components. Meanwhile, the top 10 genes in biological processes, cellular component, and molecular function were indicated. According to the annotation of COG (clusters of orthologous groups of proteins), these genes were classified into 25 different functional classes. The cluster for general function prediction (19,089, 24.77%) represented the largest group, followed by cell cycle control, cell division, chromosome partitioning (11,670, 15.14%), replication, recombination and repair (6,704; 8.70%), and unknown functions (5,286; 6.86%) [23].

The post-treatment ovarian tissue was compared with the initial ovarian tissue in the same eel. The results showed that certain pathways were significantly ($p < 0.05$) upregulated, but in the same time, the same pathways were also significantly ($p < 0.05$) downregulated. These contradictory results might imply the complexity of gene network, indeed, abundant mRNA isoforms in fish seems to play important roles on this situation (the discussion further below: 4. From ePAV to phenotypes). GSA uses all of the available gene expression data (cutoff-free) instead of pre-filtering (cutoff-based) the data for a list of strong DEGs. Our cutoff-free results indicated that most of the sequenced genes were responsible for fundamental biological regulation and metabolism. Anyways, significant metabolic activities corresponded with the change in ovarian status is indicated [23].

REViGO (http://revigo.irb.hr) was used to summarize lists of GO terms by finding a representative subset of the terms. The analysis was based on the post-treatment ovarian tissue compared with the initial ovary (e.g., 2003MT/2003). The total GO enrichment score of eels 2003, 2613, and 1440 was 184, 253, and 230, respectively. These results corresponded with the phenotypes. The outcomes from the GO enrichment analysis and phenotypes were mutually supported and appeared to have a positive correlation. The ovarian gene expression profiles from the same eel before-and after-treatment are investigated. The profiles of the heatmap or cluster indicated that eel 2613 and eel 1440 were under the same clade (cluster), and eel 2003 appeared to be isolated from both. And, the gene expression profiles were more discrete in eel 2613 between the initial ovary and post-treated ovary compared with those of the two other eels (**Figure 3**). Plotting the TMM expression of all samples and calculating the Pearson correlation coefficient, the coefficient was 0.34, 0.17, and 0.15 for eel 2003, 1440, and 2613, respectively, between the initial ovary and post-treated one in the same individual [23].

The Principal Component Analysis (PCA), the distance between any given two points represents their divergences in the chart, results are also shown: the distance between eels 1440MT and 2613MT was short on PC1 (X-axis), while eel 2003MT was isolated on the other side; eel 26130MT was closer to 2003MT while 1440MT was located from the other two on PC2 (Y-axis). The order of the intra-group distance was: 2613 > 1440 > 2003; the major shift was along PC1, and there was slight shift along PC2 except for eel 2613. Moreover, the results from the heatmap also indicated that eels 2613MT and 1440MT under the same clade (cluster), but eel 2003MT was isolated from both. The results are supported mutually (**Figure 4**) [23].

We preferred to mine the differences in gene expression profiles, thoroughly as these differences might provide the bases to set up biomarkers. To maximize the differences significantly, the cutoff condition was set with the TPM of the gene as zero in the initial ovarian tissues. We attempted to filter the differences between

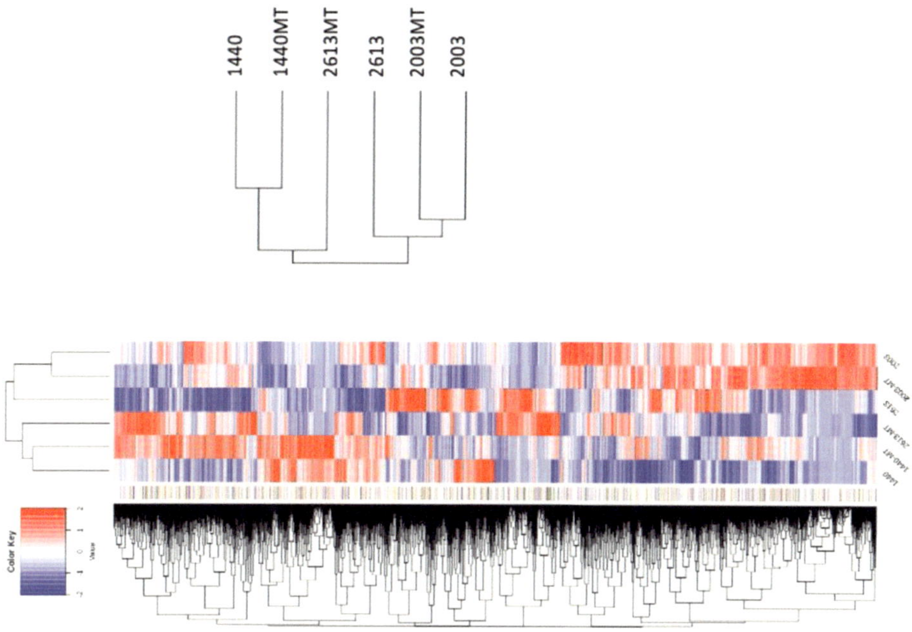

Figure 3.
Gene expression levels heatmap among different individuals. Clusters of genes with different expression values were indicated.

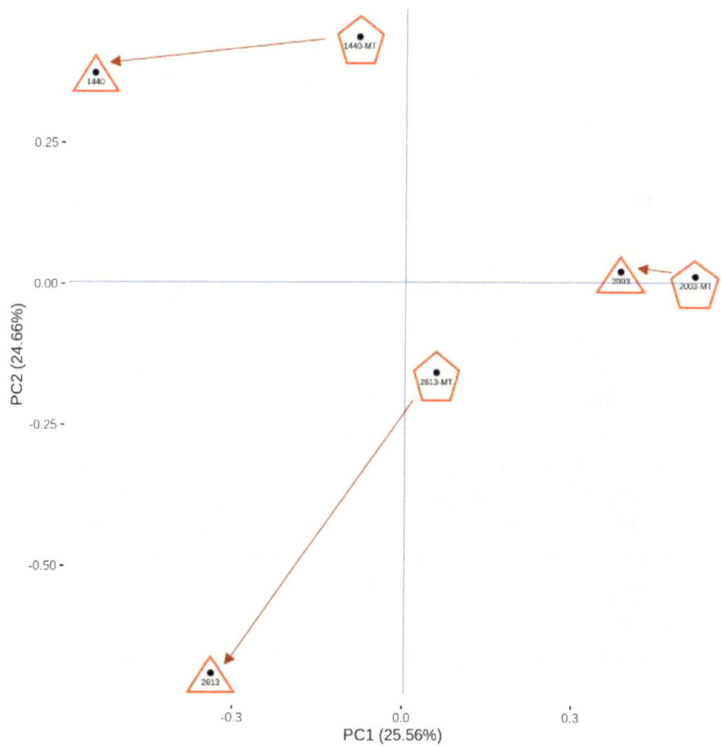

Figure 4.
Analysis gene expression profiles by PCA. The initial ovaries or the post-treated ones were symbolized as open pentagon or open triangle, respectively; the arrowhead represent the change of gene profiles.

the sensitive group (2613MT and 1440MT) and insensitive one (2003MT), the TPM was set to be equal or greater than 0.01 (TPM $\geq$ 0.01) to amplify the differences; there were 467 genes exclusively in 2613MT and 518 genes exclusively in 1440MT. As there were many isoforms, the number of unique genes was 266 and 303 for 2613MT and 1440MT, respectively. These unique genes are involved in various biological pathways. There were many GO enrichment biological pathways in addition to housekeeping metabolic pathways, the neurogenesis pathways (axon guidance, eye morphogenesis, and axon development in 2613MT; sensory organ development, regulation of synapse organization, and regulation of nervous system development in 1440MT) appeared to overlap in these two eels. These data implied plausible roles of neurogenesis in the primary development of the ovary [23].

The cutoff levels were TPM $\geq$ 0.01 and TPM = 0 for the sensitive group and insensitive group, respectively, tried to filter data based on DEGs to maximize the differences. Our assumption was that the significant ($\geq$ 0.01 *vs* 0.00) DEGs among the initial ovaries might be the basis for the outcomes of 17MT subsequently, and certain subtle pathways might be amplified; this might be arbitrary and abstract, since it is impossible to interpret the information from a transcriptome without the context of experiment.

Interestingly, several pathways implicated in neurogenesis or neuro-activity have been mined. We changed the cutoff-levels; curiously, the pathways implicating eyes or sensory organ development were found. The retina is the most accessible part of the vertebrate central nervous system, and the neural retina constitutes an excellent system with which to analyze key aspects of neurogenesis [24]. Moreover, it is well recognized that the silver eel (the prepubertal stage) has larger eyes, and a similar situation might occur in the developing gonads. Our cutoff-based results led us to postulate that neurogenesis and/or neuro-activity is involved in primary ovarian development in Japanese eel. This hypothesis might pertain to the fact that in eel, even entering the silver stage, the reproductive endocrine system is blocked, but a slight development of ovary is observed.

Few reports have discussed the involvement of neurogenesis in primary ovarian development in fish. Meanwhile, in mammals, the development of ovarian nerves precedes the onset of folliculogenesis [25], and the neural activity might be an important factor in the regulation of follicular development before the ovary acquires responsiveness to gonadotropins [26], the developing ovary is regulated by direct sympathetic inputs which function in addition but complementarily to gonadotropin [27]. On the other hand, the density of innervation may contribute to the selection of follicles for further development (reviewed by [28]). Evidence supports complementary control performed by sympathetic nerves innervating the ovary [29, 30]. Moreover, our non-*de novo* data (the sequence reads were compared with a reference genome and assembled) showed that the expression levels of ovarian doublecortin (DCX), an important marker of neurogenesis for comparative neurobiologists [31]. In summary, ovarian development in all vertebrates appears to be under at least partial neural control [32], and the neural–ovarian connection established during postnatal development is necessary [33].

Indeed, our non-*de novo* data indicated that the expression levels of ovarian *FSHR* were higher in eel 2003MT, but that in 1440MT and 2613MT was below detectable levels; it is interesting that the fold-change order of *FSHR* expression between the initial ovary and the post ovary was 2613 > 1440 > 2003. These results correspond with a stimulatory effect of androgens on ovarian *FSHR* expression levels in eels [34] or in mice [35]. This data also indicated the importance of pair-sample design in genetic studies.

Biomarkers are biological characteristics that predict treatment responses. The sensitivity of oocytes to 17MT is dependent on the ovarian status (not the mean size of the oocyte) [5]. From our results, it seems not ideal to select certain genes based on DEGs to serve as biomarkers, instead of independent genes with significant DEGs, a set of genes in certain pathways to represent the ovarian status seems chosen. The initial ovarian status is fundamental to influencing how genotypes are translated into phenotypes by treatments. It is already useful to inspect the size of sampled oocytes, while our report indicated that, the gene expression pattern can ameliorate and improve the method based on phenotype for the eel artificial maturation. Our results also implied that the neurogenesis or neural activity in the initial ovary might be an important factor influencing the outcomes of hormonal treatments. Meanwhile, more experiments and data are required to support and verify this hypothesis.

4. From ePAV to phenotypes

The term "pan-genome" is the entirety of a species gene repertoire, it is defined as the complete gene set across strains rather than in a single individual. The concept of pan-genome is built upon the observation that genes often display presence-absence variations (PAV) [36]. Pan-genome comprises of the "core" genes share by all strains within a given species, "dispensable" or "accessory" genes are only present in some strains, "unique" genes are strain-specific [36]. PAV is an important source of genetic divergence and diversity [37] and has been suggested to contribute to phenotypic variation of agronomic importance in various crops [38]. With the completion of reference genome sequencing, our knowledge of genomic variation increased, it is apparent that a single reference sequence is insufficient to represent the extent of genomic variation, resulting in the introduction and growth of the pan-genome concept [39]. In nowadays, PAV has been commonly observed across the Tree of Life [39].

The pan-transcriptome can be defined by recalling the concept of the pan-genome. It reflects the set of all the RNA present in a specie or in a single organism, transcripts present in every individual of a taxonomic group are called core genes, while transcripts absent in at least a single individual are called accessory (dispensable) genes [40]. Indeed, a larger part of the genes are expression presence-absence variation (ePAV) or are even totally absent (genomic PAV; gPAV) [41]. PAV at the genomic level would be reflected in the transcriptome ePAV. The ePAV not only reflect genomic structural variation, but also the variations in genetic and epigenetic regulatory elements [41].

As a correlation between the initial ovarian status and the phenotypic outcomes of 17MT treatment have been demonstrated [5], and the genetic background of initial ovary seems important [23]. We ameliorated our results that the ovarian transcriptome was re-mapped against the Japanese genomic scaffolds, a pair as the unit to re-analyze DEGs. Certain genes were absent in one pair but present in the others, the presence genes were also up- or down-regulated, the genes of various sets were analyzed for KEGG pathway enrichment. Together up, we speculated the existence of presence-absence variant (PAV) or expression presence-absence variation (ePAV) in Japanese eel.

The 22199 pan-ovarian transcripts were derived from ovarian tissues of 12 eels with different maturation stages. In a previous report, the number of total trinity "genes" was 336,973 [23], while there were 22,199 pan-ovarian transcripts after re-mapped against the Japanese genomic scaffolds. This significant discrepancy shows the importance of specific genomic scaffolds on transcriptomics. The number of functional genes was countered under the criterion that a gene was detected in at least one of six samples (three pairs), and the gene was eliminated when this gene was not detected in any of the six samples. The number of core genes was 15466;

there were 980, 501, and 1097 accessory genes for the intersection between 1440s and 2003s, between 2003s and 2613 s, and between 2613 s and 1440s, respectively. The numbers of unique genes were 1335, 796, and 724 for the 1440s, 2003s, and 2613 s, respectively (**Figure 5**). we emphasize PAVs here. We cannot confirm that the PAVs were stage-specific or inducible and would be detectable (expressed) in the later stages, since the ovaries of this stage remained far from maturation.

Regarding the phenotype, there was no significant difference in mean ovarian diameter among the initial ovarian tissues (**Figure 1**) [5], but all genotype data from both PCA and heatmaps indicate a closer distance between 1440MT (the initial ovarian tissue) and 2613MT, while 2003MT was more isolated from others (**Figures 3** and **4**) [23], these observations imply that there were genetic variations. Accordingly, we speculate that the expression of the absent genes may be up-regulated (from zero to detectable) by the treatment or in the following stage; if so, these absent genes are not presence-absence variations. It has been demonstrated that, not all absent genes were developmental- or stage-specific, which indirectly supports the concept of ePAV in Japanese eel [42].

It is essential to characterize the ePAV genes and their possible functions [41], the core genes has been speculated to appear universally over-represented by house-keeping functions and essential metabolic processes for organisms, and accessory genes are often associated with communication, virulence, and defense responses ([39], references therein] or adaptive functions in plants (e.g., stress responses [43]). It has also been demonstrated that accessory genes exhibit higher rates of polymorphism than core genes [44, 45]. We studied possible functions of the unique genes of PAVs in each pair by KEGG pathway enrichment testing, since phenotypes are determined or controlled by constellations of pathways. The list of unique presence genes of each pair was submitted to Molas (http://molas.iis.sinica.edu.tw/jpeel/) to test for the KEGG pathway enrichment, and the pathways were marked when the enrichment p-value <0.05; these pathways were clustered into seven categories in the KEGG pathway database. Indeed, only approximately 36% of unique genes were hit in the KEGG pathways (the hit rates were 37%, 36%, and 36% for 1440s, 2003s and 2613 s, respectively). These results indicate that the functions of transcripts in eels remained largely unknown and non-annotated, while the results also indicated that 1440s was clearly more similar to 2003.

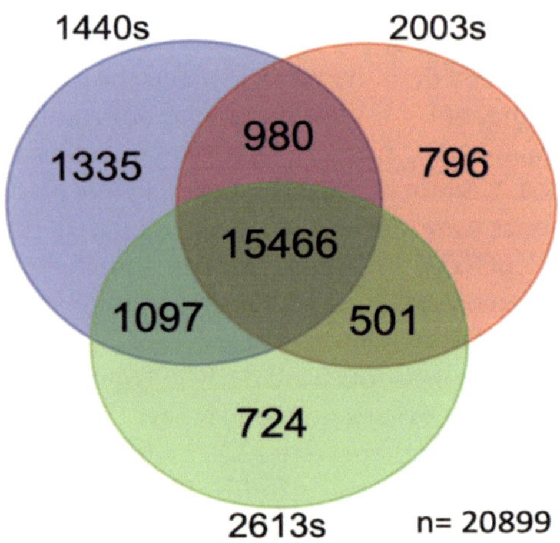

Figure 5.
Number of functional genes in each pair. Functional genes were detected in at least one sample; if not, this gene was eliminated (MT ≠ 0 ∪ AT≠0).

All signaling pathways belong to the category of "Environmental Information Processing" in KEGG, and the numbers of pathways of each pair in this category were 12, 7, and 5 for 1440s, 2003s and 2613, respectively. There were 3 pathways at the intersection of those 3 pairs: Rap1 signaling pathway, cGMP-PKG signaling pathway, and cAMP signaling pathway. This result implies the importance of these signaling pathways in basal ovarian development, and the importance of differential pathways in ovarian development was shown. Moreover, we believe that the differentially expressed genes are integrated into certain pathways to determine or control phenotypes. It has been demonstrated that the functions of genes related to neuronal activities or neurogenesis appears important in the outcomes of androgen treatments [23]. This postulation is supported the existing of ePAV in Japanese eel.

5. Conclusion

Our results, from Japanese eel, implied that the initial ovary status might be an important factor influencing the outcomes of androgen treatment, and the genes related with neuronal activities or neurogenesis seemed to play an essential role in the positive effect; certain genes seemed absent in one individual but present in the others by speculating that there is a presence-absence variant (PAV) or expression presence-absence variation (ePAV) in Japanese eel, a primitive fish diverged from other bony fishes in the basal lineage of the Teleost. Most pathways involved by ePAV were belong the endocrine system and nervous system. These results signify the importance of genetic differential expression on the variations of phenotypes by androgen, and a transcriptomic approach appears to enable extracting multiple layers of genomic data. More elegant experimental designs fit the biological model, and more transcriptomic data are required to address this question in the future.

Acknowledgements

We would like to thank Mr. Yung-Lin Huang and Mr. Yi-Sen Lai, their encouragements were appreciated sincerely. Mr. Guan-Ru Chen (Freshwater Aquaculture Research Center, Fisheries Research Institute, C.O.A., Taiwan) is also appreciated for his help on the manipulation. This research was partly funded by the Ministry of Science and Technology (MOST107-2321-B002-057), Taiwan.

Funding

This study was partly funded by the Ministry of Science and Technology (MOST107–2321-B002–057), Taiwan.

Conflict of interest

The authors declare that they have no conflict of interest.

Code availability

All software application employed are online open access.

Author details

Yung-Sen Huang[1*] and Chung-Yen Lin[2]

1 Department of Life Science, National University of Kaohsiung, Kaohsiung, Taiwan

2 Institute of Information Science, Academia Sinica, Taipei, Taiwan

*Address all correspondence to: yshuang@nuk.edu.tw

References

[1] Tanaka H, Kagawa H, Ohta H, Unuma T, Nomura K. The first production of glass eel in captivity: fish reproductive physiology facilitates great progress in aquaculture. Fish Physiol Biochem. 2003;28:493-497.

[2] Okamura A, Horie N, Mikawa N, Yamada Y, Tsukamoto K. Recent advances in artificial production of glass eels for conservation of anguillid eel populations. Ecol Freshw Fish. 2014;23:95-110.

[3] Higuchi M, Mekuchi M, Hano T, Imaizumi H. Trans-omics analyses revealed differences in hormonal and nutritional status between wild and cultured female Japanese eel (*Anguilla japonica*). PLoS One. 2019;14(5): e0209063. DOI: org/10.1371/journal.pone.0209063.

[4] Ijiri S, Kayaba T, Takeda N, Tachiki H, Adachi S, Yamauchi K. Pretreatment reproductive stage and oocyte development induced by salmon pituitary homogenate in the Japanese eel Anguilla japonica. Fish Sci. 1998; 64(4):531-537.

[5] Huang Y-S, Wu X-H, Huang P-S, Chen G-R. Correlation between the ovarian status and the androgen sensibility in the cultured Japanese eel, *Anguilla japonica*. Fish Physiol Biochem. 2020; 46:1063-1074. DOI: 10.1007/s10695-020-00772-1.

[6] Kazeto Y, Ijiri S, Todo T, Adachi S, Yamauchi K. Molecular cloning and characterization of Japanese eel ovarian P450c17 (CYP17) cDNA. Gen Comp Endocrinol. 2000;118:123-133.

[7] Walters KA, Middleton LJ, Joseph SR, Hazra R, Jimenez M, Simanainen U, Allan CM, Handelsman DJ. Targeted loss of androgen receptor signaling in murine granulosa cells of preantral and antral follicles causes female subfertility. Biol Reprod. 2012;87(151):1-11.

[8] Monson C, Forsgren K, Goetz G, Harding L, Swanson P, Young G. A teleost androgen promotes development of primary ovarian follicles in coho salmon and rapidly alters the ovarian transcriptome. Biol Reprod.2012; 97(5):731-745. DOI:10.1093/biolre/iox124.

[9] Huang YS, Chen YM, Liao PC, Lee YH, Gwo JC, Chen MC, Chang CF. Testosterone improves the transition of primary oocytes in artificial maturation eels (Anguilla japonica) by altering ovarian PTEN expression. Fish Physiol Biochem. 2012;38(3):777-787. DOI:10.1007/s10695-011-9560-6.

[10] Lokman PM, Wylie MJ, Downes M, Di Biase A, Damsteegt EL. Artificial induction of maturation in female silver eels, Anguilla australis: the benefits of androgen pre-treatment. Aquaculture. 2015;437:111-119.

[11] Mordenti O, Emmanuele P, Casalini A, Lokman PM, Zaccaroni A, Di Biase A, Parmeggiani A. Effect of aromatable androgen (17-methyltestosterone) on induced maturation of silver European eels (Anguilla Anguilla): oocyte performance and synchronization. Aquac Res.2018;49:442-448.

[12] Ijiri S, Tsukamoto K, Seinen Chow S, Kurogi H, Adachi S, Tanaka H. Controlled reproduction in the Japanese eel (Anguilla japonica), past and present. Aquac Europe. 2011;36(2): 13-17.

[13] Thorstad EB, Økland F, Westerberg H, Aarestrup K, Metcalfe JD. Evaluation of surgical implantation of electronic tags in European eel and effects of different suture materials. Mar Freshw Res. 2013;64:324-331.

[14] Wand MP, Jones MC. Kernel smoothing. Chapman & Hall/CRC monographs on statistics & applied probability (60). Boca Raton, FL, U.S.: Chapman & Hall; 1994.

[15] Tsukamoto K, Chow S, Otake T, Kurogi H, Mochioka N, Miller MJ, Aoyama J, Kimura S, Watanabe S, Yoshinaga T, Shinoda A, Kuroki M, Oya M, Watanabe T, Hata K, Ijiri S, Kazeto Y, Nomura K, Tanaka H. Oceanic spawning ecology of freshwater eels in the western North Pacific. Nature Communications. 2011;2(1):179. DOI:10.1038/ncomms1174.

[16] Tanaka S, Okamura A, Mikawa N, Yamada Y, Horie N, Utoh T, Zhang H, Oka HP, Tsukamoto K. Development of oocytes in a post-spawning Japanese eel *Anguilla japonica* Temminck & Schlegel in captivity. bioRxiv. 2018. DOI:10.1101/494435.

[17] Hsueh AJ, Kawamura K, Cheng Y, Fauser BC. Intraovarian control of early folliculogenesis. Endocr Rev. 2015;36:1-24. DOI: 10.1210/er.2014-1020.

[18] Zhang Y, Yan Z, Qin Q, Nisenblat V, Chang H-M, Yu Y, Wang T, Lu C, Yang M, Yang S, Yao Y, Zhu X, Xia X, Dang Y, Ren Y, Yuan P, Li R, Liu P, Guo H, Han J, He H, Zhang K, Wang Y, Wu Y, Li M, Qiao J, Yan J, Yan L. Transcriptome landscape of human folliculogenesis reveals oocyte and granulosa cell interactions. Mol Cell. 2018;72:1021-1034. DOI: 10.1016/j.molcel.2018.10.029.

[19] Kallen A, Polotsky AJ, Johnson J. Untapped reserves: controlling primordial follicle growth activation. Trends Mol Med. 2018;24:319-331. DOI:10.1016/j.molmed.2018.01.008.

[20] Väremo L, Nielsen J, Nookaew I. Enriching the gene set analysis of genome-wide data by incorporating directionality of gene expression and combining statistical hypotheses and methods. Nucleic Acids Res. 2013;41:4378-4391. DOI: 10.1093/nar/gkt111.

[21] Zhu B, Pardeshi L, Chen Y, Ge W. Transcriptomic analysis for differentially expressed genes in ovarian follicle activation in the zebrafish. Front Endocrinol (lausanne). 2018;9:593. DOI: /10.3389/fendo.2018.00593.

[22] Lee SC, Hsiao CD, Wang YS, Lou SW. Transcriptome response of previtellogenic ovary in Anguilla japonica after artificial hormone injection. Mar Genomics. 2017;35:31-34. DOI: 10.1016/j.margen.2017.04.002.

[23] Huang Y-S, Cheng W-C, Lin C-Y. Androgenic Sensitivities and Ovarian Gene Expression Profiles Prior to Treatment in Japanese Eel (Anguilla japonica). Mar Biotechnol (NY). 2021;23(3):430-444. doi: 10.1007/s10126-021-10035-6.

[24] Centanin L, Wittbrodt J. Retinal Neurogenesis Development. 2014;141:241-244. DOI: 10.1242/dev.083642.

[25] Malamed S, Gibney JA, Ojeda SR. Ovarian innervation develops before initiation of folliculogenesis in the rat. Cell Tissue Res. 1992; 270:87-93. DOI: 10.1007/BF00381883.

[26] Mayerhofer A, Dissen GA, Costa ME, Ojeda SR. A role for neurotransmitters in early follicular development: induction of functional follicle-stimulating hormone receptors in newly formed follicles of the rat ovary. Endocrinology. 1997;138:3320-3329. DOI: 10.1210/endo.138.8.5335.

[27] Ojeda SR, Lara HE. Role of the sympathetic nervous system in the regulation of ovarian function. In "The menstrual cycle and its disorders". Eds by Pirke KM, Wolfgang W, Ulrich S, Springer-Verlag Berlin Heidelberg. 1989.

[28] Dissen GA, Dees WL, Ojeda SR. Neural and neurotrophic control of ovarian development. In "The ovary". Eds by Adashi EY. and Leung PCK. Raven Press. New York. pp. 1-19. 1993.

[29] Lara HE, Ferruz JL, Luza S, Bustamante DA, Borges Y, Ojeda SR. Activation of ovarian sympathetic nerves in polycystic ovary syndrome. Endocrinology. 1993;133:2690-2695. DOI: 10.1210/endo.133.6.7902268.

[30] Lara HE, Dorfman M, Venegas M, Luza SM, Luna SL, Mayerhofer A, Guimaraes MA, Rosa E, Silvam AA, Ramírez VD. Changes in sympathetic nerve activity of the mammalian ovary during a normal estrous cycle and in polycystic ovary syndrome: Studies on norepinephrine release. Microsc Res Tech. 2002;59:495-450.DOI: 10.1002/jemt.10229.

[31] Balthazart J, Ball GF. Doublecortin is a highly valuable endogenous marker of adult neurogenesis in canaries. Brain Behav Evol. 2014;84:1-4. DOI: 10.1159/000362917.

[32] Nakamura M, Specker JL, Nagahama Y. Innervation of steroid-producing cells in the ovary of Tilapia Oreochromis niloticus. Zoolog Sci. 1996;13:603-608. DOI: 10.2108/i0289-0003-13-4-603.

[33] Garrido MP, Fernandois D, Venegas M, Paredes AH. Effects of sympathectomy on ovarian follicular development and steroid secretion. Reproduction. 2018;155:173-181. DOI:10.1530/REP-17-0318.

[34] Lee S-C, Lou S-W. Androgenic Modulation in the Primary Ovarian Growth of the Japanese eel, *Anguilla japonica*. Zool Stud. 2019;58:e2. DOI:10.6620/ZS.2019.58-02.

[35] Sen A, Prizant H, Light A, Biswas A, Hayes E, Lee HJ, Barad D, Gleicher N, Hammes SR. Androgens regulate ovarian follicular development by increasing follicle stimulating hormone receptor and microRNA-125 expression. Proc Natl Acad Sci U S A. 2014;111: 3008-3013. DOI: 10.1073/pnas.1318978111.

[36] Tettelin T, Masignani V, Cieslewicz MJ, Donati C, Medini D, Ward NL, Angiuoli SV, Crabtree J, Jones AL, Durkin AS, Deboy RT, Davidsen TM, Mora M, Scarselli M, Margarit y Ros I, Peterson JD, Hauser CR, Sundaram JP, Nelson WC, Madupu R, Brinkac LM, Dodson RJ, Rosovitz MJ, Sullivan SA, Daugherty SC, Haft DH, Selengut J, Gwinn ML, Zhou L, Zafar N, Khouri H, Radune D, Dimitrov G, Watkins K, O'Connor KJ, Smith S, Utterback TR, White O, Rubens CE, Grandi G, Madoff LC, Kasper DL, Telford JL, Wessels MR, Rappuoli R, Fraser CM. Genome analysis of multiple pathogenic isolates of *Streptococcus agalactiae*: implications for the microbial "pan-genome". Proc. Natl. Acad. Sci. U. S. A. 2005;02(39): 13950-5. DOI: 10.1073/pnas.0506758102.

[37] Springer NM, Ying K, Fu Y, Ji T, Yeh C-T, Jia Y, Wu W, Richmond T, Kitzman J, Rosenbaum H, Iniguez AL, Barbazuk WB, Jeddeloh JA, Nettleton D, Schnable PS. Maize inbreds exhibit high levels of copy number variation (CNV) and presence/absence variation (PAV) in genome content. PLoS Genet. 2015;5(11):e1000734. DOI: 10.1371/journal.pgen.1000734.

[38] Weisweiler M, de Montaigu A, Ries D, Pfeifer M, Stich B. Transcriptomic and presence/absence variation in the barley genome assessed from multi-tissue mRNA sequencing and their power to predict phenotypic traits. BMC Genomics. 2019;20 (1):787. DOI: 10.1186/s12864-019-6174-3.

[39] Golicz AA, Bayer PE, Bhalla Prem L, Batley J, Edwards D. Pangenomics comes of age: From bacteria to plant and animal applications. Trends Genet.

2020;36 (2):132-145. DOI:10.1016/j.
tig.2019.11.006.

[40] Medini D, Donati C, Tettelin H, Masignani V, Rappuoli R. The microbial pan-genome. Curr. Opin. Genet. Dev. 2005;15(6):589-594. DOI: 10.1016/j. gde.2005.09.006.

[41] Jin M, Liu H, He C, Fu J, Xiao Y, Wang Y, Xie W, Wang G, Yan J. Maize pan-transcriptome provides novel insights into genome complexity and quantitative trait variation. Scientific Reports, 2016;6:18936. DOI: 10.1038/ srep18936.

[42] Huang Y-S, Lin C-Y, Cheng W-C. Investigating the transcriptomic and expression presence-absence variation exist in Japanese eel (*Anguilla japonica*), a primitive teleost. Genomics. 2021 (submitted to Marine Biotechnology).

[43] Hurgobin B, Edwards D. SNP discovery using a pangenome: Has the single reference approach become obsolete? Biology (Basel). 2017;6(1):21. DOI: 10.3390/biology6010021.

[44] Glasauer SMK, Neuhauss SCF. Whole-genome duplication in teleost fishes and its evolutionary consequences. Mol. Genet. Genomics. 2014; 289(6):1045-1060. DOI: 10.1007/ s00438-014-0889-22014.

[45] Rozenfeld C, Blanca J, Gallego V, Garcıa-Carpintero V, Herranz-Jusdado JG, Perez L, Asturiano JF, Cañizares J, Peñaranda DS. De novo European eel transcriptome provides insights into the evolutionary history of duplicated genes in teleost lineages. PLoS ONE. 2019;14 (6):e0218085. DOI:10.1371/journal. pone.0218085.

Chapter 8

Intraovarian Gestation in Viviparous Teleosts: Unique Type of Gestation among Vertebrates

Mari-Carmen Uribe, Gabino De la Rosa-Cruz, Adriana García-Alarcón and Juan Carlos Campuzano-Caballero

Abstract

The intraovarian gestation, occurring in teleosts, makes this type of reproduction a such complex and unique condition among vertebrates. This type of gestation of teleosts is expressed in special morphological and physiological characteristic where occurs the viviparity and it is an essential component in the analysis of the evolutionary process of viviparity in vertebrates. In viviparous teleosts, during embryogenesis, there are not development of Müllerian ducts, which form the oviducts in the rest of vertebrates, as a result, exclusively in teleosts, there are not oviducts and the caudal region of the ovary, the gonoduct, connects the ovary to the exterior. The lack of oviducts defines that the embryos develop into the ovary, as intraovarian gestation. The ovary forms the oocytes which may develop different type of oogenesis, according with the storage of diverse amount of yolk, variation observed corresponding to the species. The viviparous gestation is characterized by the possible intimate contact between maternal and embryonic tissues, process that permits their metabolic interchanges. So, the nutrients obtained by the embryos could be deposited in the oocyte before fertilization, contained in the yolk (lecithotrophy), and may be completed during gestation by additional provisioning from maternal tissues to the embryo (matrotrophy). Then, essential requirements for viviparity in poeciliids and goodeids are characterized by: **a)** the diversification of oogenesis, with the deposition of different amount of yolk in the oocyte; **b)** the insemination, by the transfer of sperm to the female gonoduct and their transportation from the gonoduct to the germinal region of the ovary where the follicles develop; **c)** the intrafollicular fertilization; **d)** the intraovarian gestation with the development of embryos in intrafollicular gestation (as in poeciliids), or intraluminal gestation (as in goodeids); and, **e)** the origin of embryonic nutrition may be by lecithotrophy and matrotrophy. The focus of this revision compares the general and specific structural characteristics of the viviparity occurring into the intraovarian gestation in teleosts, defining this reproductive strategy, illustrated in this review with histological material in a poeciliid, of the species *Poecilia latipinna* (Lesueur, 1821) (Poeciliidae), and in a goodeid, of the species *Xenotoca eiseni* (Rutter, 1896) (Goodeidae).

IntechOpen

Keywords: oogenesis, gonoduct, intrafollicular gestation, intraluminal gestation, lecithotrophy, matrotrophy

1. Introduction

Viviparity, the marvelous and complex reproductive strategy in which the mother retains developing eggs inside their body and give birth to a living young, occurs in all the Classes of vertebrates, except in birds. Viviparity among vertebrates makes the evolutionary first appearance in fishes, becoming essential in the understanding of this reproductive strategy [1–4]. The viviparity develops an ample diversity of features along the evolution of this type of reproduction in vertebrates, this is the case of teleosts of the families Poeciliidae and Goodeidae. The adaptations for viviparity in these species developed strategies that have been successful in a wide variety of aquatic environments [3, 5, 6]. In teleosts, during embryogenesis, there are not development of Müllerian ducts, which form the oviducts in the rest of vertebrates, as a result, exclusively in these fishes, there are not oviducts, consequently there is not uterus. Therefore, the posterior zone of the ovary of teleosts, called gonoduct, characterized by the absence of germinal cells, connects the rest of the ovary to the exterior, opening at the genital pore. The lumen of the germinal region of the ovary is continuous with the lumen of the gonoduct [1–3, 7–10]. Because of the lack of oviducts in viviparous teleosts, the development of the embryos occurs into de ovary, as an intraovarian gestation, unique in vertebrates, instead of into the uterus as occurs in the rest of viviparous vertebrates. This ovarian structure defines the characteristics of the teleosts viviparity as occurs in poeciliids and goodeids. The characteristics of the ovary during non-gestation and gestation stages are presented and compared in this study, taking as models two species: the poeciliid *Poecilia latipinna* (Lesueur, 1821), and the goodeid *Xenotoca eiseni* (Rutter, 1896). During non-gestation, ovaries during previtellogenesis and vitellogenesis where selected; and, during gestation, ovaries in early, middle, and advanced gestation where selected. The ovaries were prepared for histological analysis, stained with Hematoxiline-Eosine and the image 3A is PAS [11]. Digital photomicrographs were taken using an Olympus digital camera (model C5050Z) coupled to an Olympus CX31 microscope.

The gonadal differentiation includes the initial development of two ovaries which fuse forming a single and saccular ovary with a central lumen [1, 2]. The lack of Müllerian ducts determines the lack of oviducal development. Consequently, the female reproductive system is formed only by the ovary where the gonoduct, a non-germinal caudal portion of the ovary, connects it with the exterior [8–10]. In agreement with these structural characteristics, gestation occurs in the ovary. The embryonic development in poeciliid is intrafollicular therefore, there is not ovulation, and in goodeids is intraluminal, being the embryos discharged from the follicle to the ovarian lumen just after fertilization.

Poeciliids produce eggs generally larger than goodeids. The eggs of the most of poeciliid species attains approximately 2.0–2.5 mm in diameter, meanwhile the eggs of the most of goodeid species attains approximately 0.5–1.0 mm in diameter. Similar data for species of both families are related to time of gestation period, approximately 45–60 days of gestation, and the total length of embryos at birth about 13-20 mm total length. Then, the larger eggs of poeciliid eggs compared with smaller eggs of goodeid eggs, but similar size in the newborns indicates the more intense lecithotrophy in poeciliids compared with more intense matrotrophy in goodeids. [1–3, 6]. Consequently, goodeid embryos required to absorb more nutrients derived from maternal tissues during gestation.

2. Structure of the ovary in poeciliids and goodeids

As in most viviparous teleosts, in all poeciliids and goodeids, such as *P. latipinna*, and *X. eiseni*, the ovary is a single organ, longitudinally situated. The single ovary presents a saccular structure, with a central lumen, corresponding to the cyst-ovarian type. The single ovary is the result of the fusion of both ovaries during embryological development [2, 3, 12, 13]. The ovary is surrounded by the coelom. The ovary is located dorsally to the digestive tract and remains attached to the dorsal wall of the body by the mesovarium. The ovarian wall contains four tissue layers, they are: **a)** the germinal epithelium which lines the ovarian lumen; **b)** the stroma, subjacent to the epithelium, formed by loose vascularized connective tissue, containing follicles in different stages of development; **c)** smooth muscle layers in circular and longitudinal disposition; and **d)** serosa, formed by scarce connective tissue and mesothelium at the periphery [7, 13–17]. The ovarian mucosa forms irregular folds, that extend into the ovarian lumen, called lamellae. The lamellae are lined by germinal epithelium, where follicles in diverse stages of oogenesis may be located. The germinal epithelium contains oogonia, situated among somatic epithelial cells [3, 13, 17, 18]. The internal position of the germinal epithelium in the saccular ovary defines that the ovulation in oviparous teleosts occurs into the lumen, instead of into the coelom as happens in all other vertebrates [2, 3, 13, 16, 19].

The ovary of *P. latipinna*, and *X. eiseni*, is essentially characterized by the occurrence of the oogenesis, complex and vital process of the female germinal cells, as in all animals, during their development to mature oocytes [20]. The oogenesis in viviparous teleosts, is similar to that described in oviparous teleosts, involving cyclical sequence of morphophysiological changes in the differentiation of oogonia into full-grown oocytes. According to the annual cycles, the ovary may contain only early stages of oogenesis (previtellogenesis) (**Figure 1A** and **2A**), or late stages of oogenesis (vitellogenesis) (**Figure 1B** and **2B**). Several authors analyze this process [7, 14–16, 21–24]. Oogonia is the earliest stage of germ cells which grow from an initial diameter of approximately 10 μm. Oogonia initiate meiosis becoming oocytes which are surrounded by follicular cells, developing the primordial follicles. This process comprises two main stages in sequence: previtellogenesis (**Figure 1A, C, D** and **2A, C, D**), and vitellogenesis (**Figure 3A–D** and **4A–D**). During previtellogenesis, the nucleus of the oocyte, call germinal vesicle, has a nucleolus which proliferates to multiple nucleoli, and the ooplasm initiates the growths involving a great increase in the number of organelles associated with synthetic activities, as ribosomes, endoplasmic reticulum, mitochondria, and Golgi apparatus, becoming basophilic for the staining affinity of the ooplasm. Posteriorly, during the vitellogenesis, the oocyte growth considerably when accumulates lipid globules and yolk, becoming acidophilic for the staining affinity of the yolk. Yolk is the fundamental and more abundant material stored during oogenesis for the metabolic activities required during the embryonic development. In addition to yolk, glycogen granules and lipid globules are also stored in the oocyte, forming an essential complex for embryonic nutrition. In the full-grown oocyte, the germinal vesicle moves towards the animal pole. In both, goodeids and poeciliids, during oocyte maturation the hydration of yolk occurs and becomes fluid and homogeneous, and some lipid globules may remain around the oocyte periphery [5, 7, 13].

There is a clear difference between the egg size diameter of mature eggs of poeciliids and goodeids, been larger the eggs of poeciliids. In most poeciliid species there is a mean diameter of mature eggs of 1.5 mm to 2.5 mm [6, 25–27]. Even there

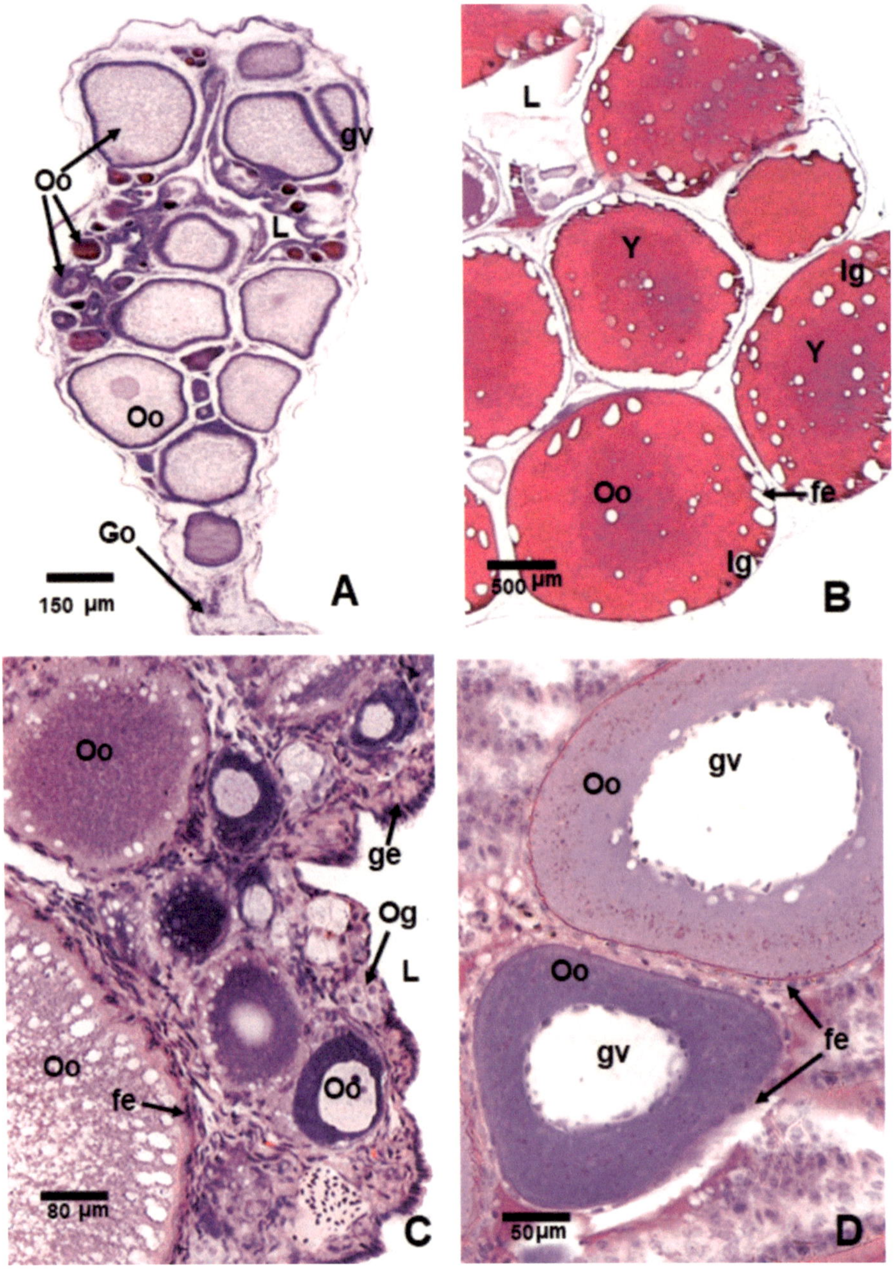

Figure 1.
Ovarian structure of Poecilia latipinna, *comparing ovaries during non-gestation with previtellogenic oocytes (A) and vitellogenic oocytes (B). A) Saccular ovary with a central lumen. Ovary during non-gestation with numerous follicles during different stages of previtellogenesis seen by the change in the oocyte diameter. In some oocytes, the nucleus (germinal vesicle) is seen. The follicular cells surround the oocytes. In the caudal region of the ovary the initial portion of the gonoduct is seen. B) Saccular ovary with a central lumen. Ovary during non-gestation with follicles during late vitellogenesis containing abundant yolk fluid and homogeneous. Lipid globules are seen at the periphery. The follicular cells surround the oocytes. C) Follicles during different stages of previtellogenesis, follicular cells surround the oocytes, a group of oogonia is seen. The lumen is surrounded by the germinal epithelium. D) Two follicles in advanced stage of previtellogenesis, the basophilia diminishes with the growth of the oocyte. (fe) follicular epithelium, (ge) germinal epithelium, (gv) germinal vesicle, (go) gonoduct, (L) ovarian lumen, (lg) lipid globules, (Og) oogonia, (Oo) oocytes, (Y) yolk. Staining technique: A-D: Hematoxiline-Eosine.*

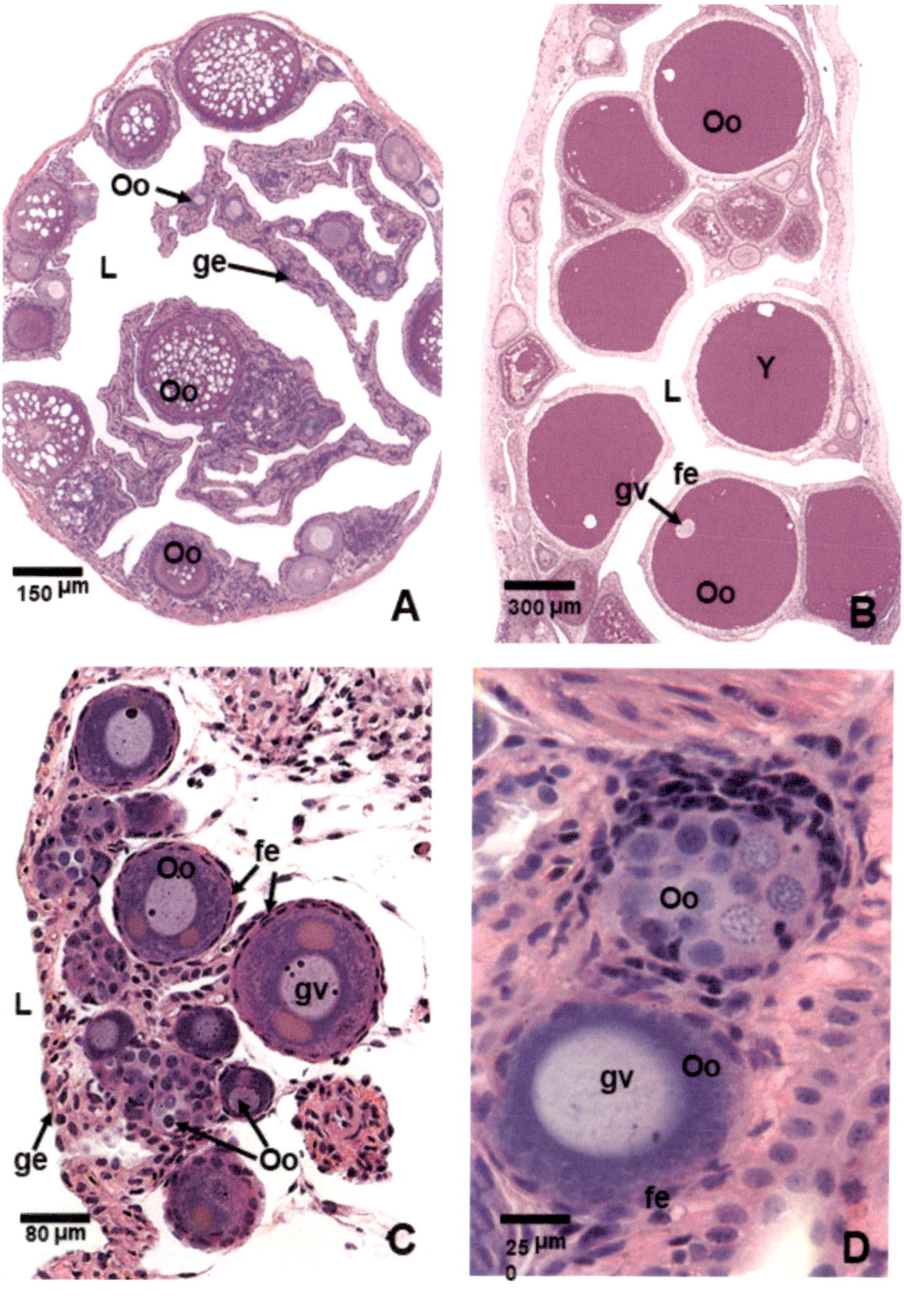

Figure 2.
Ovarian structure of Xenotoca eiseni, *comparing ovaries during non-gestation with previtellogenic oocytes (A) and vitellogenic oocytes (B). A) Saccular ovary with a central lumen. Ovary during non-gestation with follicles during different stages of previtellogenesis seen by the change in the oocyte diameter. B) Saccular ovary with a central lumen. Ovary during non-gestation with follicles during late vitellogenesis containing abundant yolk. In some oocytes, the nucleus (germinal vesicle) is seen. The follicular cells surround the oocytes. C) Follicles with oocytes in different stages of previtellogenesis, follicular cells surround the oocytes. The lumen is surrounded by the germinal epithelium. D) Detail of oocytes in early stages of previtellogenesis. The nucleus (germinal vesicle) is seen. The follicular cells surround the oocytes. (fe) follicular epithelium, (ge) germinal epithelium, (gv) germinal vesicle, (L) ovarian lumen, (Oo) oocytes, (Y) yolk. Staining technique: A-D: Hematoxiline-Eosine.*

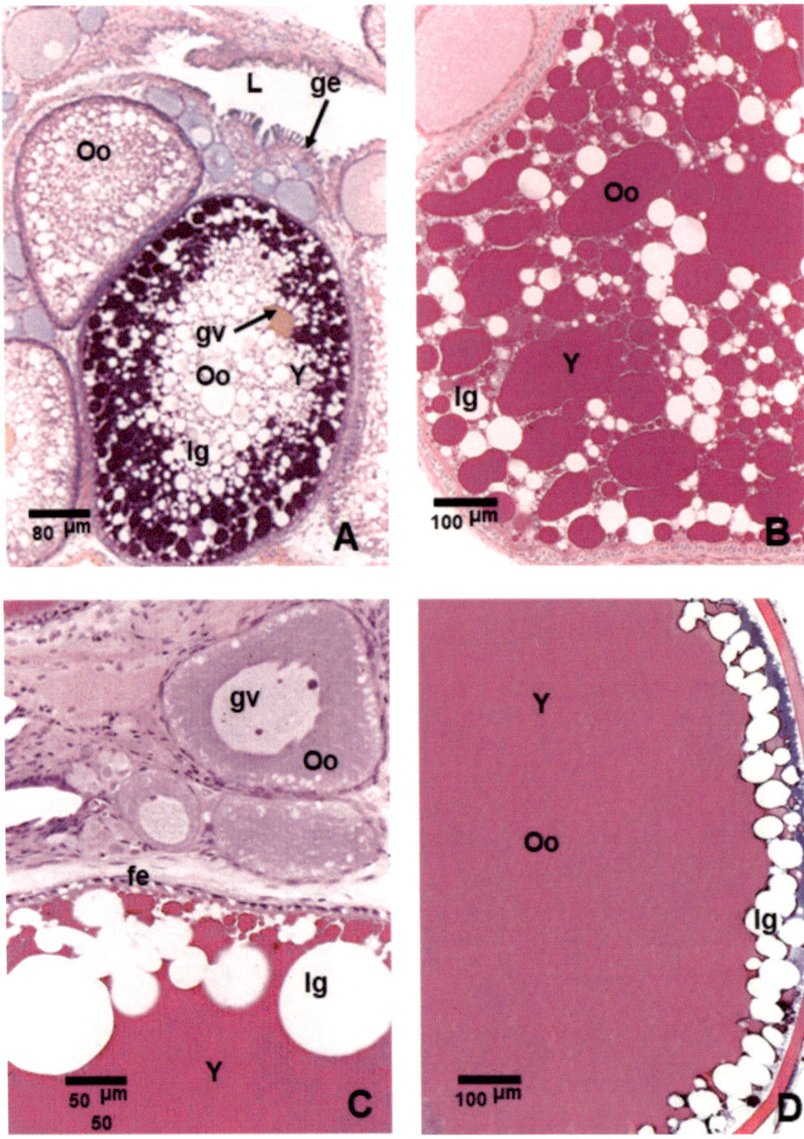

Figure 3.
Ovary of Poecilia latipinna. *A) Ovary during non-gestation with follicles during the initial vitellogenesis, with yolk platelets at the periphery. Lipid globules are seen among the yolk. The acidophilia is clear by the presence of the yolk. The ovarian lumen is surrounded by the germinal epithelium. B) Follicle in advanced stage of vitellogenesis, the yolk is initially becoming fluid and homogeneous. Abundant lipid globules are seen. C) Portion of a vitellogenic oocyte, with lipid globules at the periphery, the yolk is fluid and homogeneous. The follicular cells surround the oocytes. Compare this vitellogenic oocyte with the previtellogenic oocyte seen in the upper part of the image. D) Late vitellogenic oocyte, the yolk is completely fluid and homogeneous, some lipid globules may be seen around the oocyte periphery. (fe) follicular epithelium, (ge) germinal epithelium, (gv) germinal vesicle, (L) ovarian lumen, (lg) lipid globules, (Oo) oocytes, (Y) yolk. Staining technique: A: PAS; B-D: Hematoxiline-Eosine.*

are some poeciliids with smaller oocytes as *Heterandria formosa*, which mature oocyte attains an egg diameter of 0.4 mm, considered the smallest egg of poeciliids [28]. The species *P. latipinna* forms mature eggs with a mean diameter of 2.2 mm [26]. In most goodeid the mature egg has a mean diameter of 0.6 mm to 1.0 mm., the species *X. eiseni* forms mature eggs with a mean diameter of about 0.8 mm in diameter [29].

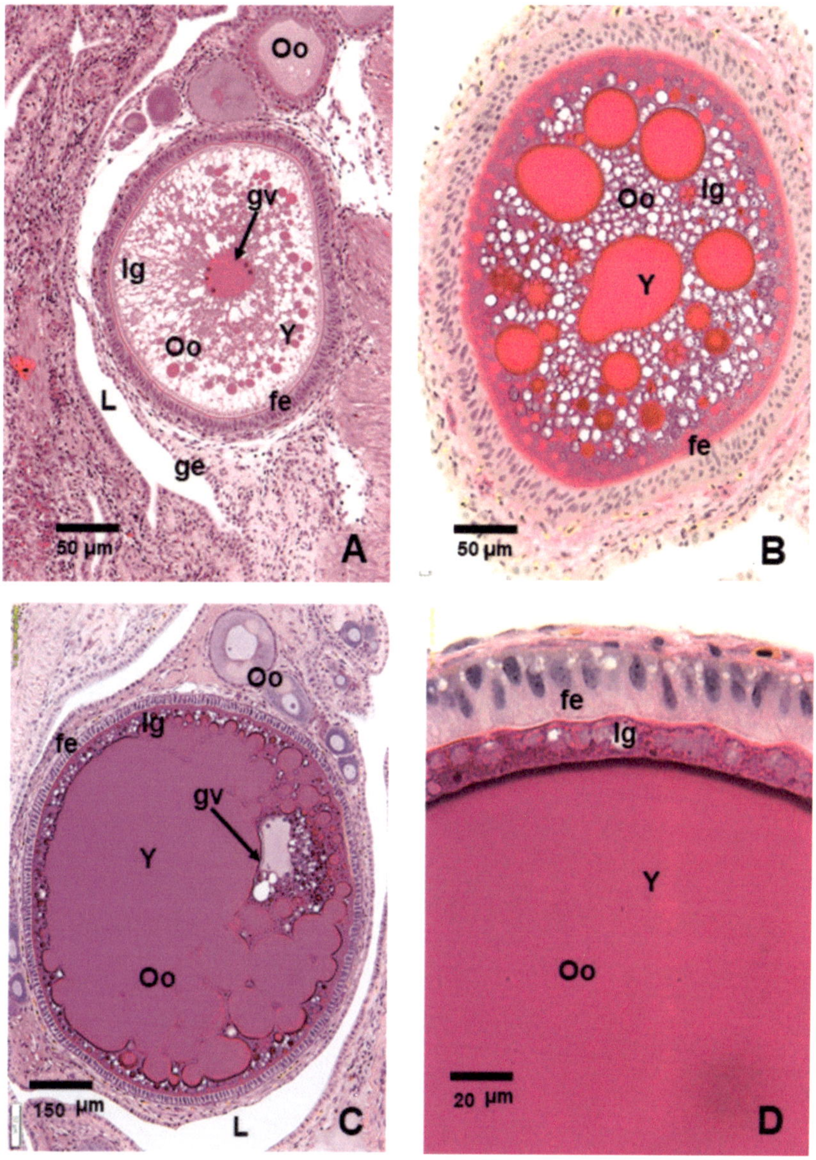

Figure 4.
Ovary of Xenotoca eiseni. *A) Ovary during non-gestation with follicles in the initial vitellogenesis, with yolk platelets at the periphery. Lipid globules are seen among the yolk. The follicular cells surround the oocytes. The acidophilia is clear by the presence of the yolk. The ovarian lumen is surrounded by the germinal epithelium. B) Follicle in middle stage of vitellogenesis, the yolk is initially becoming fluid and homogeneous. Abundant lipid globules are seen. The follicular cells surround the oocyte. C) Late vitellogenic oocyte, with lipid globules at the periphery, the yolk is fluid and homogeneous. The follicular cells surround the oocytes. The nucleus (germinal vesicle) is seen at the animal pole. Compare this vitellogenic oocyte with the previtellogenic oocytes seen in the upper part of the image. D) Late vitellogenic oocyte, the yolk is completely fluid and homogeneous, some lipid globules may be seen around the oocyte periphery. (fe) follicular epithelium, (ge) germinal epithelium, (gv) germinal vesicle, (L) ovarian lumen, (lg) lipid globules (Oo) oocytes, (Y) yolk. Staining technique: A-D: Hematoxiline-Eosine.*

3. Fertilization

Viviparity is basically related to the evolution of reproductive behavior, insemination and fertilization are essential requirements for viviparity. In viviparous species, as poeciliids and goodeids, the internal fertilization involves the insemination, with the entrance of spermatozoa to the female reproductive system, through the transfer of sperm from the male to the female gonoduct, and the sequence of the entrance of the sperms, from the gonoduct to the germinal zone of the ovary, where occurs the internal fertilization [1, 19, 30–32]. The secretory activity of the ovarian epithelium, having a trophic function for the spermatozoa, contributes for their survival during the appropriate time for the fertilization. In *P. latipinna*, and *X. eiseni*, the spermatozoa may be seen in the ovarian lumen (**Figure 5A** and **6A**), many of them show their heads in contact with the apical end of the germinal epithelial cells.

Insemination in viviparity may allow the temporal separation of mating and fertilization, and consequently, the temporal diversification of the time

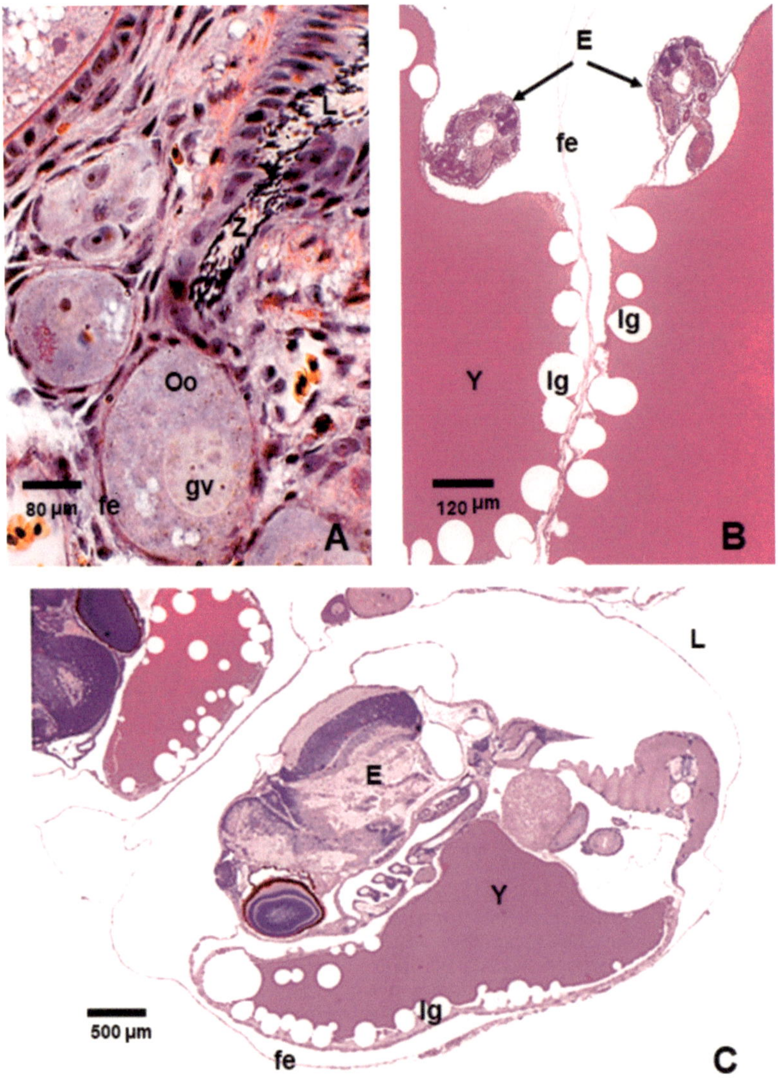

Figure 5.
Ovary of Poecilia latipinna. *A) When insemination occurred, the ovary contains abundant spermatozoa in the lumen. Several previtellogenic oocytes are seen, surrounded by follicular epithelium. B) Ovary during intrafollicular gestation with embryos in early stage of development, after neural tube formation, with abundant yolk. Lipid globules are seen around the yolk. The follicular epithelium surrounds the follicle in gestation. C) with the advance of the gestation, the embryo growths and the yolk reduce. Lipid globules continue at the periphery. The follicular epithelium surrounds the follicle in gestation. (E) Embryo, (fe) follicular epithelium, (gv) germinal vesicle, (L) ovarian lumen, (lg) lipid globules, (Oo) oocytes, (Y) yolk, (Z) spermatozoa. Staining technique: A-C, Hematoxiline-Eosine.*

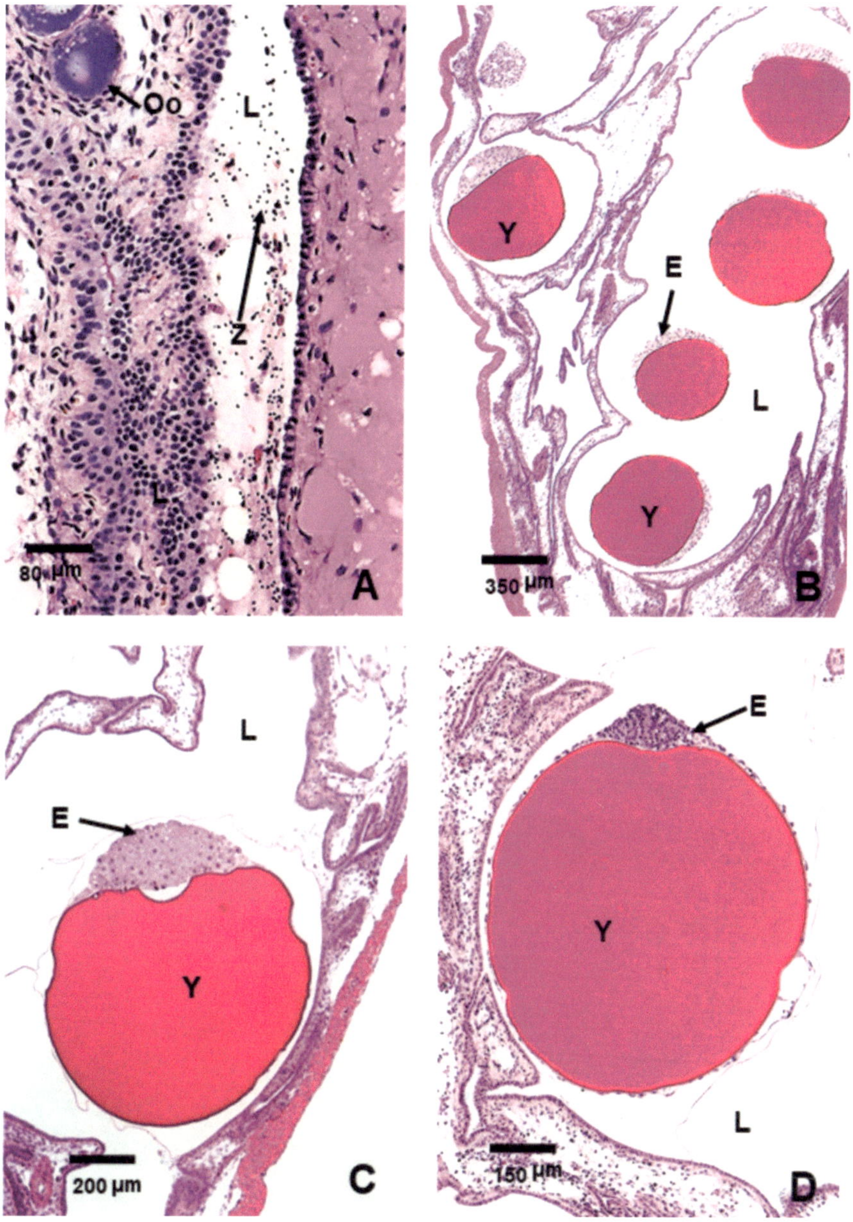

Figure 6.
Ovary of Xenotoca eiseni*. A) When insemination occurred, the ovary contains abundant spermatozoa in the lumen. A previtellogenic oocyte is seen. B) Ovary during intraluminal gestation with embryos in the lumen during blastula, with abundant yolk. C) Detail of an embryo during blastula in the ovarian lumen. D) Embryo in intraluminal gestation in early stage of development, after neural tube formation, with abundant yolk. (E) Embryo, (L) ovarian lumen, (Oo) oocyte, (Y) yolk, (Z) spermatozoa. Staining technique: A-D: Hematoxiline-Eosine.*

between insemination and birth, by the possibility of sperm storage in the ovary. Additionally, the sperm storage permits that, with one insemination, several clutches of eggs maturing on several occasions, may be fecundated [31].

In most viviparous teleosts, as poeciliids and goodeids, fertilization occurs inside the follicle, as an intrafollicular fertilization, therefore, there is not ovulation. During the fertilization, sperm must penetrate the follicular wall to reaches the oocyte membrane [31, 33–35].

4. Intrafollicular and intraluminal gestation in poeciliids and goodeids

After the intrafollicular fertilization the embryogenesis may follows two different ways of gestation: **a)** intrafollicular gestation, where the embryos remain in the follicle during their development and move into the ovarian lumen immediately before birth, as occurs in poeciliids; or **b)** intraluminal gestation, in which the embryos, during early development as blastula, move from the follicle to the ovarian lumen where gestation continues until birth, as occurs in goodeids, in this type of gestation the embryos in the lumen are surrounded by a fluid, call histotrophe, secreted by the

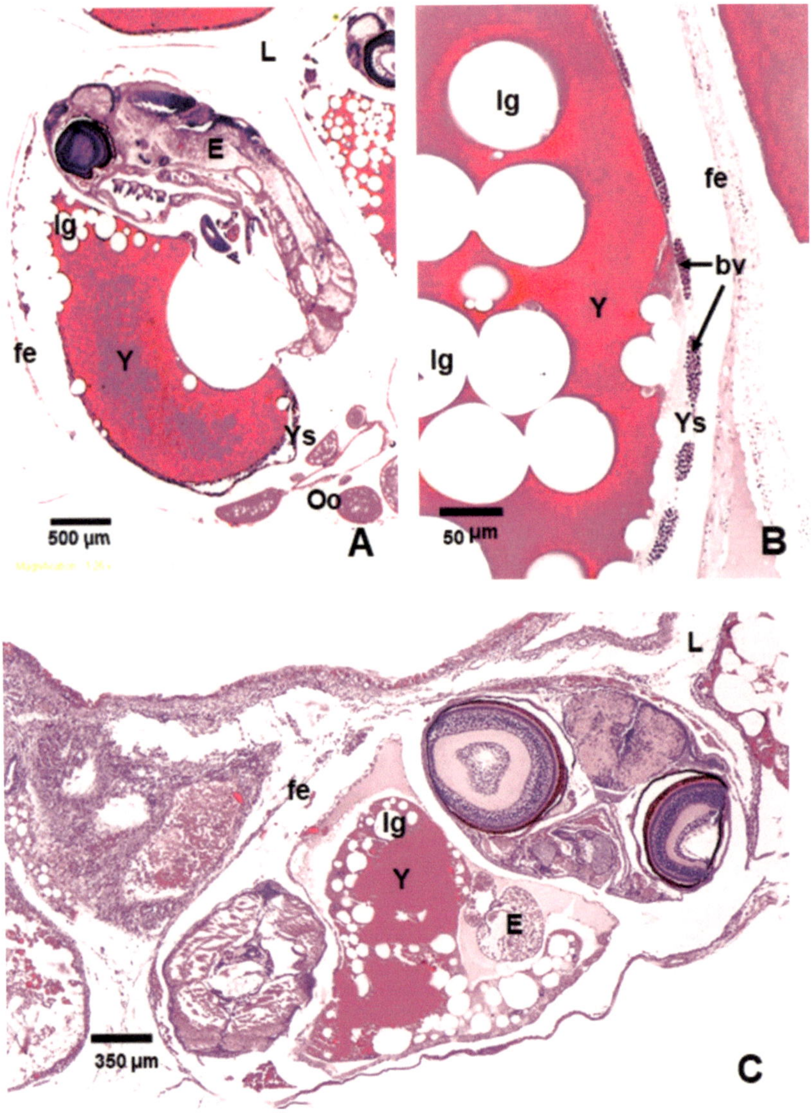

Figure 7.
Ovary of Poecilia latipinna *in intrafollicular gestation. A) Ovary during gestation with an embryo in middle stage of development. Lipid globules are seen at the periphery. Around the yolk is seen the yolk sac. The follicular epithelium surrounds the follicle in gestation. Oocytes are seen in the ovarian wall. B) Periphery of the yolk, where it is seen the vascularized yolk sac with blood vessels. Lipid globules are seen. The follicular epithelium surrounds the follicle in gestation. C) an embryo during advanced stage of gestation, the amount of yolk diminishes with the growth of the embryo. Lipid globules are seen. The follicular epithelium surrounds the follicle in gestation. (bv) blood vessels, (E) embryo, (fe) follicular epithelium, (L) ovarian lumen, (lg) lipid globules, (Oo) oocytes, (Y) yolk, (Ys) yolk sac. Staining technique: A-C, Hematoxiline-Eosine.*

ovarian epithelium to the lumen [1, 29, 30, 36, 37]. Thus, the intrafollicular gestation is the case of the poeciliid *P. latipinna*, and the intraluminal gestation is the case of the goodeid *X. eiseni*. Embryos of *P. latipinna* in early intrafollicular gestation (**Figure 5B** and **C**) and embryos of *X. eiseni*, in early intraluminal gestation (**Figure 6B–D**), are seen. Embryos of *P. latipinna* in late intrafollicular gestation (**Figure 7A–C**) and embryos of *X. eiseni*, in late intraluminal gestation (**Figure 8A–C**), are seen.

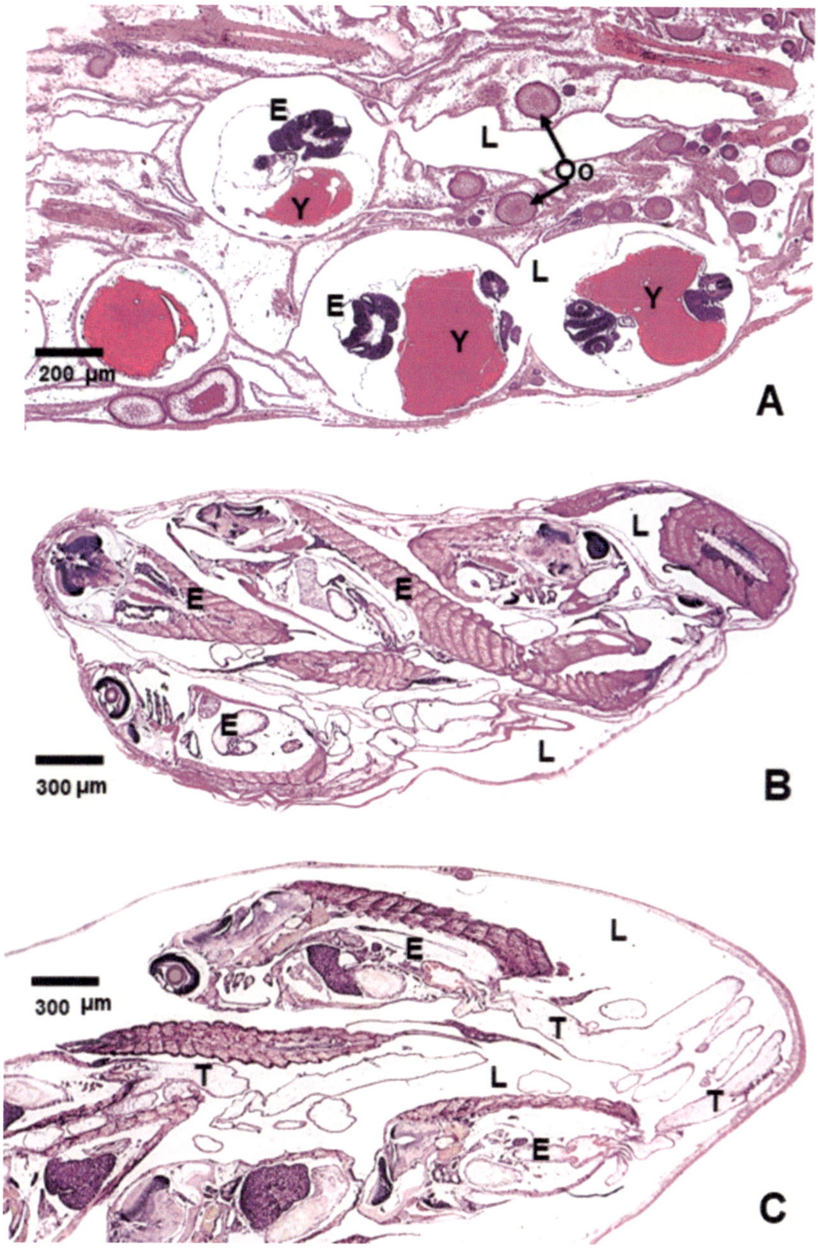

Figure 8.
Ovary of Xenotoca eiseni *in intraluminal gestation. A) Ovary during gestation with embryos in the lumen during middle stage of development, with early development of eyes. Reduced yolk in the ventral side of the embryo. Oocytes are seen in the ovarian wall. B) Ovary with embryos in the lumen during advanced stage of development. There is not yolk. C) Ovary with embryos during advanced stage of gestation. The trophotaeniae is seen as extensions of the intestine to the ovarian lumen. (E) Embryo, (L) ovarian lumen, (Oo) oocytes, (T) trophotaeniae (Y) yolk. Staining technique: A-C: Hematoxiline-Eosine.*

Consequently, both types of gestation, intrafollicular and intraluminal, define different relationships between the tissues of the embryos and the mother and successively, diverse adaptations for metabolic interchanges between them, reflected mainly in the type of embryonic nutrition. Nutrition during gestation in viviparous species reflects essential strategies which impact in the evolution of vertebrate viviparity, as occurs in goodeids and poeciliids [19, 38, 39].

In recognition of vertebrate diversity, we follow Blackburn & Starck [40] in using the term "embryo" in the broad sense, to include all types of developing offspring in gestation.

5. Embryonic nutrition by lecithotrophy and matrotrophy

The maternal–embryonic relationships include trophic, osmoregulatory, excretory, respiratory, endocrinological, and immunological processes [4, 36, 37, 40]. These interchanges are essentially related to the diversity of mechanisms of embryonic nutrition. During intraovarian gestation of poeciliids and goodeids, two types of embryonic nutrition may occur: **a)** lecithotrophy, wherein the nutrients are provided by the yolk stored in the oocyte during oogenesis; and, **b)** matrotrophy, where the nutrients are transferred by maternal tissues to the embryo during gestation. Both types of nutrition are not mutually excluding, may occurs both with different degree of nutrient transfer. These strategies require a modification in the timing of maternal resource offered to the embryo, these are: during oogenesis, previously to fertilization, in lecithotrophy, and posteriorly to fertilization, in matrotrophy. Several studies analyze the adaptations from lecithotrophy to matrotrophy [4, 19, 27, 29, 38, 40–44].

In lecithotrophic species, the yolk develops a big and vascularized yolk sac around the yolk, in the ventral part of the embryo (**Figure 7A** and **B**). The yolk sac has a simple epithelium and a plexus with abundant capillaries, structures that allow the reception of the nutrients from the yolk and their distribution to the rest of the embryonic body, throughout the circulation. During gestation, the yolk sac progressively diminishes (**Figure 7C**), as the yolk is used for the nutrition of the embryo [19, 45, 46].

The viviparity generates the opportunity for the embryo to acquire resources from the mother, developing specialized structures for interchanges, as it is the development of placental structures [27, 42, 47]. In the intrafollicular gestation, maternal tissues (follicular epithelium and the subjacent net of capillaries), and embryonic structures establish the maternal–embryonic relationships during gestation [3, 19].

According to the differences of amount of yolk, the species having eggs with abundant yolk are lecithotrophic, even some transfer of maternal nutrients may also occur through the contact of the maternal blood vessels subjacent to the follicular epithelium, which surrounds the embryo in development, as occurs in poeciliids, as *P. latipinna* (**Figure 5B** and **C**). In contrast, the species having eggs with scarce yolk deposited during oogenesis, observed during early gestation in goodeids as *X. eiseni* (**Figure 6B–D** and **8A**), but, when the scarce yolk is early consumed, and the embryo needs nutrients from the mother for the continuation of the development are matrotrophic (**Figure 8B** and **C**).

Lecithotrophy has been revealed to be better adapted energetically to seasonally unpredictable habitats, whereas matrotrophy requires a predictable food supply during gestation. Therefore, matrotrophy requires stable conditions throughout gestation for a constant transfer of nutrients from the mother to the embryos. Then, the condition of the habitat determines the convenience of lecithotrophy or matrotrophy [42, 47–49].

Early in the embryogenesis of goodeids, when the yolk is completely consumed (**Figure 8B** and **C**), the embryonic growth is progressively dependent of maternal nutrients through matrotrophy, initially by the nutrients dissolved in the histotrophe and then, by the transfer of nutrients via the development of complex and diverse placental structures where occurs the apposition of embryonic and maternal tissues [29, 36, 39, 43, 50–52]. Several embryonic structures have been considered for the transfer of nutrients from the histotrophe such as: the skin, the gills, the digestive tract through the mouth, and the characteristic structure of goodeids, the trophotaeniae [1, 7, 29, 39, 52–54]. The trophotaeniae is seen as extensions of the intestine to the ovarian lumen (**Figure 8C**). Additionally, transfer of nutrients occurs by placental structures, such as: the apposition of trophotaeniae to maternal epithelium, or by branquial placenta, a structure formed when folds of ovarian tissue enter the embryonic branchial chamber developing apposition of the ovarian folds to the embryonic branquial epithelium [49, 52, 55].

Embryos developing in the ovarian lumen, as *X. eiseni* (**Figure 6C**, **D** and **8A–C**), are bathed by histotrophe, a fluid rich in nutrients containing abundant proteins derived from the maternal blood, secreted by the ovarian tissues, and discharged in the lumen [29, 55]. The absorption of these nutrients is reached by goodeid embryos through long and vascularized extensions of the hindgut that grow outside of the embryo into the ovarian lumen developing extraordinary absorptive structures call trophotaeniae (**Figure 8C**). Trophotaeniae is also involved in gas exchange and develop process of immunity [19, 29, 30]. This capacity of absorption is evidenced by the morphological features of the trophotaeniae, as the surrounding columnar epithelium with apical brush-border and the abundant vascularization subjacent to the epithelium [29]. Then through the epithelium, the blood vessels of the trophotaeniae transport the contains from the histotrophe to the embryo. Numerous morphological and experimental studies of the trophotaeniae analyzed the transfer of nutrient to the embryonic body, as: [29, 39, 50, 52–54]. Ecdysis of the trophotaeniae occurs at birth; when the postnatal phase initiates the oral nutrition [19, 30, 36, 53].

In poeciliids during birth, the embryos are going from the follicle to the ovarian lumen and then through the gonoduct to the exterior. In goodeids during birth, the embryos are already in the ovarian lumen, then through the gonoduct they move to the exterior.

6. Conclusions

The structure of the ovary and the lack of oviducts of teleosts define the characteristics of the viviparity with the intraovarian gestation, as it was presented morphologically in the sequence of Figures of ovarian histology of the poeciliid *P. latipinna*, and the goodeid *X. eiseni* in this analysis.

The ovary of viviparous teleosts is not only the organ where occurs the oogenesis, but also it receives the spermatozoa during insemination, occurs the fertilization of oocytes and the embryos remain throughout their development until birth, making the intraovarian gestation of viviparous teleosts a such complex and unique type of gestation in vertebrates.

The oogenesis is similar to that described in oviparous teleosts, since the formation of the primordial follicle, integrated by the oocyte surrounded by the follicular epithelium, and through previtellogenesis and vitellogenesis. In accordance with the species, as it is seen in poeciliids and goodeids, the oogenesis organizes oocytes with different diameters corresponding to the amount of yolk.

The intraovarian gestation may be as: **a)** intrafollicular, where the embryos remain in the follicle during their development and move into the ovarian lumen immediately before birth, or **b)** intraluminal, in which the embryos, during early development move from the follicle to the ovarian lumen where gestation continues until birth.

The origin of embryonic nutrition occurs by lecithotrophy and matrotrophy developing active structures in the transfer of nutrients to the blood vessels of the embryo. The lecithotrophy, occurring in the intrafollicular gestation of *P. latipinna*, develops the yolk sac, which decrease progressively during gestation. The matrotrophy occurring in the intraluminal gestation of *X. eiseni,* allows the provision of nutrients from the histotrophe, and develop the trophotaeniae and the branchial placenta.

Acknowledgements

The authors are grateful to Harry J. Grier (In Memoriam) for valuable insightful reflections about this theme.

Author details

Mari-Carmen Uribe*, Gabino De la Rosa-Cruz, Adriana García-Alarcón and Juan Carlos Campuzano-Caballero
Facultad de Ciencias, Laboratorio de Biología de la Reproducción Animal, Departamento de Biología Comparada, Universidad Nacional Autónoma de México, Ciudad de México, México

*Address all correspondence to: mari3uribe3@gmail.com

References

[1] Turner CL. Viviparity in teleost fishes. Scientific Monthly. 1947; *65*, 508-518.

[2] Wourms JP. Viviparity: The maternal-fetal relationship in fishes. American Zoologists. 1981, *21,* 473-515.

[3] Uribe MC, Grier HJ, De la Rosa-Cruz G, García-Alarcón A. Chapter 3. Modifications in ovarian and testicular morphology associated with viviparity in teleosts. In: BGM Jamieson (Ed.). Reproductive biology and phylogeny of fishes (Agnathans and Bony Fishes). 2009. (Vol. 8A, pp. 85-117). Science Publishers.

[4] Blackburn DG. Evolution of vertebrate viviparity and specializations for fetal nutrition: a quantitative and qualitative analysis. Journal of Morphology. 2015. *276*, 961-990.

[5] Parenti Lynne R, Grier HJ. Evolution and phylogeny of gonad morphology in bony fishes. Integrative and Comparative Biology. 2004. 44 (5):333-348.

[6] Ponce de León JL, Rodríguez R, Acosta M, Uribe MC. Egg size and its relationship with fecundity, newborn length, and female size in Cuban poeciliid fishes (Teleostei: Cyprinodontiformes). Ecology of Freshwater Fish. 2011. 20, 243-250.

[7] Uribe MC, De la Rosa Cruz G, García Alarcón A, Campuzano-Caballero JC, Guzmán Bárcenas MG. Structures associated with oogenesis and embryonic development during intraovarian gestation in viviparous teleosts (Poeciliidae). Fishes. 2019. *4*(2), 35.

[8] Campuzano-Caballero JC, Uribe MC. Structure of the female gonoduct of the viviparous teleost *Poecilia reticulata* (Poeciliidae) during non-gestation and gestation stages. Journal of Morphology. 2014. *275*, 247-257.

[9] Campuzano-Caballero JC, Uribe MC. Functional morphology of the gonoduct of the viviparous teleost *Poeciliopsis gracilis* (Heckel, 1848) (Poeciliidae). Journal of Morphology. 2017. *278*, 1647-1655.

[10] Santamaría-Martín C, Plaul SE, Campuzano-Caballero JC, Uribe MC, Barbeito CG. Structure of the gonoduct of the viviparous teleost *Cnesterodon decemmaculatus* (Jenyns, 1842) (Poeciliidae). Journal of Morphology. 2021. 282, 533-542.

[11] Humason GL. Animal Tissue Techniques. W. H. Freeman. 1979.

[12] Dodd TM. The structure of the ovary in nonmammalian vertebrates. In: Zuckerman S, Weir BJ (Eds.), The Ovary, 2nd Ed. (Vol. 1, pp. 219-263). Academic Press. 1977.

[13] Grier HJ, Uribe MC, Patiño R. The ovary, folliculogenesis, and oogenesis in teleosts. In: B. Jamieson (Ed.) Reproductive biology and phylogeny of fishes (Agnathans and Bony Fishes). Science Publishers. 2009. pp 25-84.

[14] Wallace RA, Selman K. Cellular and dynamic aspects of oocyte growth in teleosts. American Zoologists. 1981. 21: 325-343.

[15] Selman K, Wallace RA. Cellular aspects of oocyte growth in teleosts. Zoological Science. 1989. 6: 211-231.

[16] Grier HJ, Uribe MC, Parenti LR, De la Rosa-Cruz G. Fecundity, the germinal epithelium and folliculogenesis in viviparous fishes. In: Uribe MC, Grier HJ (Eds). Viviparous Fishes New Life Publications. 2005. (pp. 191-126).

[17] Grier HJ, Uribe MC, Lo Nostro FL, Mims SD, Parenti LR. Conserved form

and function of the germinal epithelium through 500 million years of vertebrate evolution. Journal of Morphology. 2016. 277: 1014-1044.

[18] Uribe MC, Grier HJ, Avila-Zúñiga SA. The ovary of the teleost fish *Xenotoca eiseni* (Goodeidae), where in addition to the oogenesis occur insemination, fertilization and gestation. JSM Invitro Fertilization. 2017. 2(3):1-4.

[19] Wourms PJ, Grove BD, Lombardi J. The maternal-embryonic relationship in viviparous fishes. in Fish. Physiology, Hoar WS, Randal DJ. (Eds). Academic Press, Inc.: New York, NY, USA, 1988. 1-134.

[20] Starck JM, Stewart JR, Blackburn DG. Phylogeny and evolutionary history of the amniote egg. Journal of Morphology. 2021. *282*, 1080-1122.

[21] Guraya SS. The cell and molecular biology of fish oogenesis. Karger. Basel, Switzerland. 1986. 20-223.

[22] Tyler CR, Sumpter JP. Oocyte growth and development in teleosts. Reviews in Fish Biology and Fisheries. 1996. 6, 287-318.

[23] Grier HJ, Uribe MC, Parenti LR. Germinal epithelium, folliculogenesis, and postovulatory follicles in ovaries of rainbow trout, *Oncorhynchus mykiss* (Walbaum, 1792) (Teleostei, Protacanthopterygii, Salmoniformes). Journal of Morphology. 2007. 268, 293-310.

[24] Lubzens E, Young G, Bobe J, Cerda J. Oogenesis in teleosts: How fish eggs are formed. General and Comparative Endocrinology. 2010. 165, 367-389.

[25] Scrimshaw NS. Egg Size in Poeciliid Fishes. Copeia. 1946(1): 20-23.

[26] Snelson Jr. FF, Wetherington JD, Large HL. The relationship between interbrood interval and yolk loading in a generalized poeciliid fish, *Poecilia latipinna*. Copeia. 1986(2): 295-304.

[27] Marsh-Matthews E, Brooks M, Deaton R, Tan H. Effects of maternal and embryo characteristics on post-fertilization provisioning in fishes of the genus *Gambusia*. Eecologia. 2005. 144, 12-24.

[28] Uribe MC, Grier HJ. Oogenesis of microlecithal oocytes in the viviparous teleost *Heterandria formosa*. Journal of Morphology. 2011. 272, 241-257.

[29] Iida A, Arai HN, Someya Y, Inokuchi M, Onuma TA, Yokoi H, Suzuki T, Hondo E, Sano, K. Mother-to-embryo vitellogenin transport in a viviparous teleost *Xenotoca eiseni*. Proceedings of the National Academy of Sciences. USA. 2019. 116(44), 22359-22365.

[30] Turner CL. The unique nutritional organs in the embryos of the top minnows of the Mexican plateau. Science. 1933. 77:93-94.

[31] Burns JR, Weitzman SH. Insemination in Ostariophysan Fishes. In Viviparous Fishes. Uribe MC, Grier HJ. (Eds). New Life Publications: Homestead, FL, USA. 2005. 105-132.

[32] Greven H. Structural and behavioral traits associated with sperm transfer in Poeciliinae. In: Uribe MC, Grier HJ. (Eds). Viviparous Fishes, New Life Publications: Homestead, FL, USA, 2005. 145-163.

[33] Burns JR, Weitzman SH, Grier HJ, Menezes NA. Internal fertilization, testis and sperm morphology in glandulocaudinae fishes (Teleostei: Characidae: Glandulocaudinae). Journal of Morphology. 1995. 224(2), 131-145.

[34] Javonillo R, Burns JR, Weitzman SH. Sperm transfer: the first step toward viviparity. In: Viviparous Fishes,

Uribe MC, Grier HJ. (Eds). New Life Publications: Homestead, FL, USA. 2005. 143-150.

[35] Uribe MC, Grier HJ, Parenti LR. Ovarian structure and oogenesis of the oviparous goodeids *Crenichthys baileyi* (Gilbert, 1893) and *Empetrichthys latos* Miller, 1948 (Teleostei, Cyprinodontiformes). Journal of Morphology. 2012. 273, 371-387.

[36] Wourms JP. Functional morphology, development, and evolution of trophotaeniae. In: Uribe MC, Grier HJ. (Eds). Viviparous Fishes New Life Publications. 2005. (pp 217-–242).

[37] Wourms JP, Lombardi J. Reflections on the evolution of piscine viviparity. American Zoologists. 1992. *32,* 276-293.

[38] Blackburn DG. Evolutionary origins of viviparity in fishes. In: Uribe MC, Grier HJ (Eds), Viviparous Fishes. New Life Publications. 2005. (pp 303-317).

[39] Schindler JF, de Vries U. Maternal embryonic relationships in the goodeid teleost *Xenoophorus captivus* - Embryonic structural adaptations of viviparity. Cell and Tissue Research. 1987. 247, 325-338.

[40] Blackburn DG, Starck JM. Morphological specializations for fetal maintenance in viviparous vertebrates: an introduction and historical retrospective. Journal of Morphology. 2015. 276, E1–E16.

[41] Blackburn DG. Convergent evolution of viviparity, matrotrophy and specializations for fetal nutrition in reptiles and other vertebrates. American Zoologists. 1992. *32*, 313-321.

[42] Pollux BJA, Reznick DN. Matrotrophy limits a female's ability to adaptively adjust offspring size and fecundity in fluctuating environments. Functional Ecology. 2011. 25(4), 747-756.

[43] Schindler JF. Structure and function of placental exchange surfaces in goodeid fishes (Teleostei: Atheriniformes). Journal of Morphology. 2015. *276*, 991-1003.

[44] Banet AI, Au AG, Reznick DN. Is mom in charge? implications of resource provisioning on the evolution of the placenta. Evolution. 2010. 64-11, 3172-3182.

[45] Marsh-Matthews E, Deaton R, Brooks M. Survey of matrotrophy in lecithotrophic poeciliids. In: Viviparous Fishes II. Uribe MC, Grier HJ (Eds). New Life Publications: Homestead, FL, USA. 2010. 255-258.

[46] Ponce de León JL, Uribe MC. Morphology of yolk and pericardial sacs in lecithotrophic and matrotrophic nutrition in poeciliid fishes. Journal of Morphology. 2021. *282*, 887-899.

[47] Pollux BJA, Pires MN, Banet AI, Reznick DN. Evolution of placentas in the fish family Poeciliidae: An empirical study of macroevolution. Annual Review of Ecology, Evolution, and Systematics. 2009. 40, 271-289.

[48] Reznick DN, Callahan H, Llauredo R. Maternal effects on offspring quality in poeciliid fishes. American Zoologists. 1996. 36, 147-156.

[49] Turner CL. Pseudoamnion, pseudochorion, and follicular pseudoplacenta in poeciliid fishes. Journal of Morphology. 1940. 67, 59-89.

[50] Mendoza G. The ovary and anal processes of *"Characodon" eiseni,* a viviparous cyprinodont teleost from Mexico. Biological Bulletin. 1965. *129*, 303-315.

[51] Schindler JF, Hamlett WC. Maternal-embryonic relations in viviparous teleosts. Journal of Experimental Zoology. 1993. 266, 378-393.

[52] Uribe MC, Grier HJ, Avila-Zúñiga SA, García-Alarcón A. Change of lecithotrophic to matrotrophic nutrition during gestation in the viviparous teleost *Xenotoca eiseni* (Goodeidae). Journal of Morphology. 2018. 279(9): 1336-1345.

[53] Iida A, Nishimaki T, Sehara-Fujisawa A. Prenatal regression of the trophotaenial placenta in a viviparous fish *Xenotoca eiseni*. Scientific Reports. 2015. 5: 7855.

[54] Iida A, Eiichi Hondo JN. Histological observation of the reproductive system in a viviparous teleost *Xenotoca eiseni* Rutter 1896 (Cyprinodontiformes: Goodeidae). Anatomia, Histologia, Embryologia. 2021. 50(1): 161-168.

[55] Uribe MC, De la Rosa-Cruz G. García-Alarcón A. Branchial placenta in the viviparous teleost *Ilyodon whitei* (Goodeidae). Journal of Morphology. 2014. *275,* 1406-1417.